DELIVERING EXCELLENT SER\ IN AVIATION

For Tamara, Joseph and Loveday

Delivering Excellent Service Quality in Aviation

A Practical Guide for Internal and External Service Providers

MARIO KOSSMANN

LONDON AND NEW YORK

First published 2006 by Ashgate Publishing

2 Park Square, Milton Park, Abingdon, Oxon OX14 4RN
711 Third Avenue, New York, NY 10017, USA

Routledge is an imprint of the Taylor & Francis Group, an informa business

First issued in paperback 2016

British Library Cataloguing in Publication Data
Kossmann, Mario
   Delivering excellent service quality in aviation: a
   practical guide for internal and external service providers
   1.Aeronautics, Commercial - Quality control 2.Aeronautics,
   Commercial - Management 3.Aircraft industry - Quality
   control 4.Aircraft industry - Management
   I.Title
   387.7'068

Library of Congress Cataloging-in-Publication Data
Kossmann, Mario.
   Delivering excellent service quality in aviation: a practical guide for internal
and external service providers / by Mario Kossmann.
      p. cm.
   Includes bibliographical references and index.
   ISBN 0-7546-4725-0
   1. Airplanes--Design and construction--Quality control. 2. Aircraft industry--
Customer services--Quality control. 3. Airlines--Customer services--Quality control.
4. Airports--Management--Quality control. I. Title.

   TL671.28.K67 2006
   629.134'2--dc22

                                                              2006000098

ISBN 978-0-7546-4725-6 (hbk)
ISBN 978-1-138-26256-0 (pbk)

Contents

List of Figures   *vii*
List of Tables   *ix*
About the Author   *xi*
Acknowledgements   *xii*
Glossary   *xiii*

Chapter 1 Introduction   1
    Chapter Summary   1
    Why You Should Read This Book   1
    Why Does It Pay to Improve Service Quality?   2
    Provision of Service Quality by Internal Service Providers   3
    The Market for Service Providers in the Aviation Context   6
    Preview of Operating Areas Covered   7
    How to Use This Book   11

Chapter 2 Theoretical Considerations   13
    Chapter Summary   13
    General Service Issues   13
    The Gaps Model of Service Quality   20
    The Five Performance Objectives   45
    The Importance-Performance Matrix   51
    Change Management Issues   55
    Quality in Aviation   62

Chapter 3 The Service Quality Cycle   75
    Chapter Summary   75
    Overview   75
    Step 1 – Generation of Service Standards   76
    Step 2 – Ranking of Service Standards   90
    Step 3 – Measuring Performance against Service Standards   96
    Step 4 – Analysing the Current Situation   102
    Step 5 – Deriving Action Items   111
    Step 6 – Closing the Cycle   117
    Summary of Benefits   126

Chapter 4 Case Study: Aircraft Manufacturing   129
          Chapter Summary   129
          Context of the Case Study   129
          Status Before   133
          Implementation of the Service Quality Cycle   133
          Status After   166
          Lessons Learned   169

Chapter 5 Conclusion   175

Appendix: The Service Quality Cycle Checklists   *177*

Bibliography   *187*

Index   *189*

List of Figures

Figure 1.1	Service quality spells profits	3
Figure 1.2	Service providers in the aviation context (examples)	4
Figure 1.3	Passenger traffic and capacity growth	7
Figure 1.4	Comparison of major German airports	10
Figure 1.5	How to use this book	11
Figure 2.1	The service optimum	15
Figure 2.2	The gaps model of service quality	20
Figure 2.3	Closing the Customer Gap (Gap 5)	21
Figure 2.4	Nature and determinants of customer expectations	22
Figure 2.5	Factors influencing customer perceptions of service	25
Figure 2.6	Closing Provider Gap 1	28
Figure 2.7	Closing Provider Gap 2	34
Figure 2.8	Process for setting customer-defined standards	35
Figure 2.9	Basic principle of customer integration	36
Figure 2.10	Closing Provider Gap 3	38
Figure 2.11	Closing Provider Gap 4	42
Figure 2.12	Approaches for matching service delivery and promises	43
Figure 2.13	Internal and external effects of performance objectives	50
Figure 2.14	The polar representation of performance objectives	51
Figure 2.15	The importance-performance matrix	53
Figure 2.16	Example of the applied importance-performance matrix	54
Figure 3.1	The service quality cycle	76
Figure 3.2	Generation of service standards	77
Figure 3.3	Step 1 – Generation of service standards	77
Figure 3.4	Ranking of service standards	90
Figure 3.5	Step 2 – Ranking of service standards	92
Figure 3.6	Measuring performance against service standards	98
Figure 3.7	Step 3 – Measuring performance	99
Figure 3.8	Analysing the current situation	104
Figure 3.9	Step 4 – Analysing the current situation	104
Figure 3.10	Generic importance-performance matrix	106
Figure 3.11	Deriving action items	111
Figure 3.12	Step 5 – Deriving action items	112
Figure 3.13	Closing the cycle	118
Figure 3.14	Step 6 – Closing the cycle	119

Figure 4.1   Organization of the Support Services Group   130
Figure 4.2   Organization of the service provider   131
Figure 4.3   Organization and locations of customer teams   132
Figure 4.4   Support services coverage in Germany (status before)   134
Figure 4.5   Performance against service standards, March 2004   138
Figure 4.6   Performance against service standards, May 2004   147
Figure 4.7   Tendencies, March–May 2004   148
Figure 4.8   Performance against service standards, July 2004   152
Figure 4.9   Tendencies, May–July 2004   152
Figure 4.10   Tendencies, March–July 2004   153
Figure 4.11   Performance against service standards, December 2004   158
Figure 4.12   Tendencies, March–December 2004   159
Figure 4.13   Tendencies, March–December 2004 (simplified)   159
Figure 4.14   Support services coverage in Germany (status after)   168
Figure 4.15   Budget development over time   168

List of Tables

Table 2.1	Expanded marketing mix for services	17
Table 2.2	Ways in which services marketers can influence factors	24
Table 2.3	A nine-point scale of importance	52
Table 2.4	A nine-point scale of performance	53
Table 2.5	Strategies for dealing with resistance	59
Table 2.6	The eight-stage process of creating major change	61
Table 2.7	Criteria for performance excellence	68
Table 2.8	Criteria for airline ranking	71
Table 2.9	Examples of airline rankings	72
Table 3.1	Checklist – marketing research	79
Table 3.2	Checklist – customer relationship management	81
Table 3.3	'Quality' means different things in different operations	82
Table 3.4	'Speed' means different things in different operations	83
Table 3.5	'Dependability' means different things in different operations	83
Table 3.6	'Flexibility' means different things in different operations	84
Table 3.7	Some typical partial measures of performance objectives	86
Table 3.8	Checklist – performance objectives (internal)	87
Table 3.9	Checklist – interviews and focus groups	87
Table 3.10	Checklist – customer involvement (Step 1)	89
Table 3.11	Checklist – service standards (external)	90
Table 3.12	Checklist – scale of importance	93
Table 3.13	Checklist – interviews, focus groups and questionnaires	94
Table 3.14	Checklist – customer involvement (Step 2)	96
Table 3.15	Checklist – table of ranking results	96
Table 3.16	Checklist – scale of performance	100
Table 3.17	Checklist – customer involvement (Step 3)	101
Table 3.18	Checklist – table of performance results	102
Table 3.19	Checklist – importance-performance matrix	107
Table 3.20	Checklist – situational context	108
Table 3.21	Checklist – analysis	109
Table 3.22	Checklist – strategy	113
Table 3.23	Checklist – contractual obligations	114
Table 3.24	Checklist – infrastructure	114
Table 3.25	Checklist – financial resources	115
Table 3.26	Checklist – human resources	116
Table 3.27	Checklist – action items	116
Table 3.28	Checklist – dealing with resistance	120

Table 3.29   Checklist – priority   121
Table 3.30   Checklist – timing   122
Table 3.31   Checklist – communications to customers   123
Table 3.32   Checklist – closing the cycle   124
Table 3.33   Summary of benefits   126

Table 4.1   Initial service standards (internally defined)   135
Table 4.2   Consolidated and ranked service standards (externally defined)   136
Table 4.3   Support services performance measurements questionnaire   137
Table 4.4   Analysis of the current situation, March 2004   139
Table 4.5   List of derived action items, March 2004   146
Table 4.6   Analysis of the current situation, May 2004   149
Table 4.7   List of derived action items, May 2004   151
Table 4.8   Analysis of the current situation, July 2004   154
Table 4.9   List of derived action items, July 2004   157
Table 4.10   Analysis of the current situation, December 2004   160
Table 4.11   List of derived action items, December 2004   164
Table 4.12   Lessons learned   170
Table 4.13   Costs of implementing the service quality cycle   174

About the Author

Mario Kossmann is a Systems Engineer and Capability Integrator for Airbus, having previously worked for Blohm & Voss as a Systems Engineer, Technical Manager and also Consultant in Services Marketing. He has served as an officer with the German and French navies, and was awarded a MEng in Aerospace Engineering from the University of the Federal Armed Forces in Munich, and an MBA from the University of Warwick.

In 2005, Mario transferred from Airbus Deutschland to Airbus UK, tasked with the development and implementation of an integrated service quality concept for one of the core engineering disciplines within Airbus. This concept has been validated and implemented transnationally (in the UK, Germany, France and Spain) across all aircraft development programmes and across all engineering centres of competence and excellence.

Acknowledgements

I would like to thank all those who enabled, supported, encouraged and tolerated me while I was working on this book. I would like to make particular mention of the following: Arne Sievert (Kienbaum Management Consultants) and Inge Radler (MAN Turbo) in the concept phase of writing the book; Guy Loft (Ashgate Publishing) in the refinement and delivery phases; Professor Robert Johnston and David Arnott (University of Warwick), Professor Dr. Claudia Fantapié Altobelli (Helmut Schmidt Universität Hamburg) and Professor Dr. Arnold Hermanns (Universität der Bundeswehr Munich) during the review phase; as well as my wife and children throughout the whole process – they gave me their continued support by commenting on the contents of the book, brainstorming with me on particular issues, critically discussing different aspects with me, or simply accepting me being obsessed about doing research for and actually writing the book in Germany, Spain, France and the UK over a period of two years. Also, I would like to thank Airbus, Securitas, Munich Airport, SKYTRAX, dba, UPS and Fraport AG for their kind permission to use their material.

Glossary

Terms/Abbreviations	Definition
ACMT	Aircraft Component Management Team (see Case Study).
Capability	Here, the term is defined to stand for best-practice processes, methods and tools (supporting the methods) in specific fields of engineering. One capability is likely to be of highest relevance at a certain stage in the development process, other capabilities will be most relevant at other stages of the development process.
Capability Integrator (CI)	A person who integrates one or several capabilities within engineering or development/design teams, enabling all relevant team members to use best-practice processes, methods and tools (thereby contributing to reduced development time and costs, as well as better products).
CDBT	Component Design Build Team (see Case Study).
CI	See 'Capability Integrator'.
Client	See 'Customer'.
CMIT	Component Management Integration Team (see Case Study).
Coaching	The consulting activity of teaching individual customers or entire customer teams the application of a specific capability, including process, method and/or tool issues. In contrast to training, coaching usually takes place in the working environment of the customers and is likely to cover very specific issues that are of more direct concern to attendees. Typically, coaching meetings vary between 30 minutes and 4 hours.
Customer	An individual (person), group or organization that uses or sells on services and related goods or products delivered by a service provider.
	Internal Customers inside the own organization. Depending on how a company is organized, customers will have to allocate part of their budget to the internal service provider (in the case of a cost centre organization) or even have to pay for the services just like external

customers would have to (in the case of a profit centre organization).

External
Customers outside the own organization. They would normally have to pay for the services delivered.

DBT   Design Build Team (see Case Study).

Focus group   A group of existing and/or potential customers that is invited by a service provider to talk about service-related issues. Concrete questions or just a general topic can be given to the group to discuss. The aim is to benefit from group interaction to gain deeper insight into real customer concerns and opinions. A moderator should make sure that the discussion is not overly dominated by one or several individuals in the group. The latter would have the negative effect that important viewpoints do not surface because some individuals feel intimidated and, hence, less motivated to speak up.

Helpdesk support   The support given by members of a support team via phone or e-mail to individual customers who encounter any problem to do with a specific capability, be it related to a process, method or tool.

Interview   A face-to-face discussion with an individual customer or potential customer. Usually interviews will be conducted in a structured manner; that is, questions or at least topics to cover are prepared. The advantage lies in the comparability and potential comprehensiveness inherent in this approach. The drawbacks are that if the questions are not well prepared and formulated the interviewer might not get the information needed and also group interaction is specifically excluded.

Low-cost airline   An airline that competes in the market predominantly on offering lower prices than other airlines for no-frills flights. The main emphasis of low-cost airlines' internal operations is on reducing costs; for instance, by suppressing tickets, only accepting Internet bookings, reducing turnaround times between flights (= increasing productivity) and restricting inflight service to the strict minimum.

Marketing   'Individual and organizational activities that facilitate and expedite satisfying exchange relationships in a dynamic environment through the creation, distribution, promotion and pricing of goods, services and ideas in return for something of value' (Dibb et al. 1997).

Operation	A set of business processes that often cut across functionally based micro-operations. Most operations produce some mix of tangible products and less tangible services. Only few operations produce either products or services alone.
Positioning	The process of creating an image or moving the perceived position of a service provider – in the minds of target customers – against specific dimensions and compared to competitors.
Questionnaire	A number of specific questions in written form that are sent to customers in order to obtain measurable or at least comparable feedback from a larger group of people.
Ranking	Here, ranking is defined as the customer activity of putting service standards in an order of priority or importance. The result of a ranking exercise may be that individual customers or customer teams attach different importance to specific service standards than other customers do.
Safety	Freedom from danger, harm or risk (fire, accidents, etc.).
Security	Protection against lawbreaking, violence, enemy acts, etc. (for example, bomb attempts, hi-jacking, theft, espionage).
Service	The 'production of an essentially intangible benefit, either in its own right or as a significant element of a tangible product, which through some form of exchange satisfies an identified consumer need' (Palmer 1994).
Service operation	Operation that predominantly produces less tangible services.
Service provider	An individual or a team delivering services to customers. *Internal* Services are predominantly delivered to internal customers inside the own organization. *External* Services are predominantly delivered to external customers outside the own organization.
Service standard	A clearly specified objective that a service provider aims to achieve in delivering services to its customers and that is communicated both internally to employees and externally to customers. It ought to be realistic,

	achievable, meaningful, easy to measure, and above all reflect customer needs and requirements.
Sub-contractor	Individual, group or organization – either internal or external to the organization of the customer – that fulfils contractually specified activities, needs and/or requirements in the form of goods, products and services for the customer. A sub-contractor is also a supplier but not all suppliers are sub-contractors.
Supplier	Individual, group or organization – either internal or external to the organization of the customer – that supplies goods, products and/or related services that are needed by the customer.
Support services	Here, support services are defined to be those services delivered to engineering or development/design teams that are necessary to enable them and their individual members to apply and make best use of specific capabilities (best-practice processes, methods and tools). Support services comprise training, coaching and helpdesk support.
Training	The activity of teaching a group of individual customers or entire customer teams the application of a specific capability, including process, method and tool issues. Usually, training takes place in isolation from the workplace of attendees and typically take between 4 hours and 1 day per unit.

Chapter 1

Introduction

Chapter Summary

The purpose of this chapter is to give you a brief overview of the content, context and usefulness of this book. In order to do this, the importance of controlling and improving service quality, both for external and internal service providers in the aviation industries, is highlighted. Then, the market for service providers in the aviation context, specifically within airline, airport and aircraft manufacturing operations, is discussed. Last, depending on your expectations from this book, three alternative ways of how the book could be read are suggested, so that you are only confronted with those parts of it that are most relevant to your situation.

Why You Should Read This Book

This book offers you a hands-on, step-by-step approach to service quality management as a practical guide for any service provider in the aviation industry, whilst, at the same time, containing more detailed reading about the underlying state-of-the-art theories, as well as practical experience with the suggested method in the form of a recent case study.

This book would be of interest to anybody who is delivering professional services to external or internal customers in the context of *airline, airport and aircraft manufacturing operations* (including respective suppliers and sub-contractors), and who wishes to measure and monitor (and therefore be able to control) perceived customer satisfaction with the services delivered over time. Their role could be that of brand manager, team leader, sub-unit leader or individual service provider.

This book is addressed at the same time to service providers in the context of airline, airport and aircraft manufacturing operations because all three domains are highly interdependent and have large overlap areas in their customer segments, the principal end customer segment being air traffic passengers. While this is obvious for airlines, it seems less so for aircraft manufacturers. Still, all major manufacturers are currently focusing their efforts to achieve even higher passenger orientation in designing their cabins. Similarly, many airports are going a long way in order to find out proactively what passengers want and need, and designing their operations and layouts accordingly.

The intention of this book is to be a *practical* and *realistic* guide to implementing (with minimum effort) an effective yet efficient way to control and proactively influence the service quality as perceived by the customer. The consequences of this are that users of this book who own decision-making, as well as budget and capacity planning, will be carrying out these activities on a sound basis, able to justify their decisions to any stakeholders in the service operation. Also, and most importantly, customer satisfaction can be improved considerably over time. Finally, improvements made can be systematically measured in the way suggested, which, in turn, leads to increased levels of motivation amongst an organization's own personnel. By defining, quantifying and reporting quality, it is possible to enhance employee awareness of the fact that high quality is actually achievable and trainable at little cost, but that the potential benefits of delivering high quality services are likely to be very large.

Many of the principles underlying the method that is proposed in this book could be argued to be valid also in other technical and non-technical industry sectors. However, service providers in the aviation context (with its high complexity, fierce competition, rapid and innovative service and product developments) focus on service as one of the main marketing assets, and a high degree of internationality as the second most important: both are particularly in need of effective and efficient service quality management. Therefore, the examples used in this book (and in fact the whole book) are clearly rooted in the aviation context.

To summarize, this book offers any service provider in the aviation context – specifically within airline, airport and aircraft manufacturing operations – an efficient yet effective way to increase customer satisfaction and, thereby, profits (or budget allocation for internal service providers). Also, it helps service providers to create a sound basis for operational and strategic decision-making.

Why Does It Pay to Improve Service Quality?

Although cost factors are far from being unimportant, there is overwhelming evidence that service quality is the single most important issue in running customer service operations successfully. It can be argued that service quality directly and indirectly affects profits in a significant way, as Zeithaml and Bitner (2002) suggest (see Figure 1.1). They argue that high service quality leads to customer retention, which has shown to be cheaper in the long run than high levels of customer turnover. Also, long-term customers tend to buy larger volumes and higher price premium services and products. Very importantly, 'word-of-mouth' communications are affected in a positive way, being the most influential and convincing kind of communication in the field of services. If people talk positively to other potential customers about their experiences with the services delivered, the supplier's market share is likely to grow, too. All this leads to higher possible margins. By means of more 'offensive marketing' (such as putting aggressive promotional campaigns into practice), market share can be affected, a positive reputation enhanced and the service offer can be positioned in a way to allow for premium pricing strategies. All this leads to higher

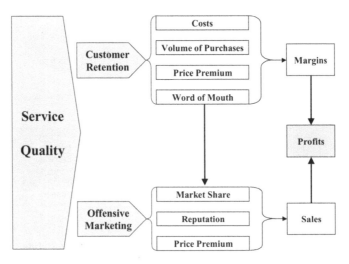

Figure 1.1   Service quality spells profits
Adapted from Zeithaml and Bitner (2002)

sales levels. Both higher sales levels and higher achievable margins directly result in increased profits (Zeithaml and Bitner 2002).

In a context where a service provider operates in a cyclical downturn of an industry or mainly serves customers that do so, the question may well be how to cope better with lower and lower budgets or sales rather than how to increase budgets or sales.

Still, whether operating in a downturn or upturn of the market, it can be said that the better the service quality delivered to customers, the better the standing of a service provider in comparison to competitors (external service provider) or the easier it is to justify and defend a specific budget requested (internal service provider).

Provision of Service Quality by Internal Service Providers

Who are the Internal Service Providers?

In order to answer this question we should first look at some typical examples of internal and external service providers in the aviation context. Figure 1.2 gives a limited yet realistic systematic overview of typical service providers involved in airline, airport or aircraft manufacturing operations. Although clearly simplifying matters, the diagram illustrates the interdependence between some of the service operations from the three aviation segments considered.

Delivering Excellent Service Quality in Aviation

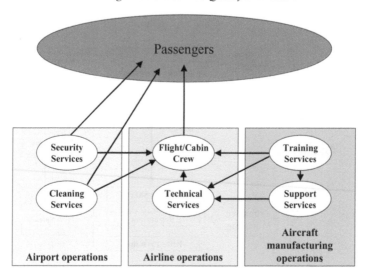

Figure 1.2   Service providers in the aviation context (examples)

The most obvious service delivery takes place from an airline's flight and cabin crew to their passengers. This external service provision to paying customers usually consists primarily of flying the passengers to their chosen destination safely and taking care of them during the journey on board.

However, in order to be able to offer that service to passengers, the flight and cabin crew need to take receipt of technical support and maintenance services so that the aircraft is fully airworthy and, for instance, inflight entertainment, galley and toilet systems are fully operational. These services can either be delivered by an internal service provider within the airline or they can be sub-contracted to an external service provider such as the technical services team of a different airline or an airport.

Both the flight crew and the technical services team must have their members trained so that they are sufficiently skilled to fly the aircraft or to maintain it respectively. An external service provider (usually the aircraft manufacturer that has produced the aircraft used by the airline) may deliver such training services.

This same service provider is likely to also offer training services internally to other teams within the aircraft manufacturing operations; for example, to the support services group that, in turn, may act as an external service provider to the airline's technical services team in cases of more complicated maintenance or repair work.

Since most passengers prefer to fly in a clean aircraft, some cleaning services will have to be delivered to the flight and cabin crew. This is usually done by an external service provider from within the local airport operations. This service provider can either be an internal part of the airport operations or sub-contracted. It can also be argued that the cleaning services team directly serves the passengers insofar as the 'personal space' of an individual customer is cleaned, but there is no direct contractual relationship.

Finally, security services from within the airport operations act as an external service provider to the airline's flight and cabin crew by protecting the aircraft on the ground and security-checking all the luggage to be taken on board as well as the passengers themselves. It can be argued that the security services are also directly delivered to passengers because the former represent protection of each passenger from physical harm through acts of violence: at many airports passengers have to pay an explicit security fee. This kind of service is very often sub-contracted to experienced security companies that meet the demanding requirements of the local authorities.

To summarize the above, it can be argued that service providers in all three aviation segments under consideration find themselves in a network of highly interdependent external and internal customer-supplier relationships. Many service providers do not directly serve their end customers but rather serve another service provider who does. In order for them to optimize and generally improve their service quality, it is very seldom sufficient to merely talk to the 'middleman' (that is, the service provider directly served); end customers and their needs and wants must be understood.

Hence, successful aircraft manufacturers not only talk to airlines but also, just as importantly, to passengers in order to find out about their requirements. Similarly, the best airports in terms of quality do not just talk to customer airlines, but also to end customers – that is, passengers.

The Importance of Providing Service Quality

Having looked at some typical service providers from the aviation context, we should now address the above question of why service quality is equally important to internal service providers. Whether you are an external or internal service provider, customers will finally make available those resources you need to run your service operation – this may be by receiving direct payments for services delivered to an external customer or by being allocated a specific budget inside a bigger company or institution. Such budget allocations are likely to happen in longer time intervals, often on a yearly basis. Still, some decision-makers in your organization will regularly look at what services you are providing internally and what added value lies in using your services. It does not come as a surprise that if your internal customers are not convinced that your service delivery makes a difference, the budgets allocated to you will at least be reduced if not suspended altogether.

If, however, you deliver *excellent* service quality:

- your internal customers will realize that you do add value to their processes and appreciate your contributions;
- the budgets necessary to effectively enhance your internal customers' activities will be allocated to you, because your internal customers realize that in paying you they actually save money and time while improving their own products;

- you give evidence that your budgets, from an internal customer's perspective, must not be considered as costs but as investments with a clear return for the internal customer.

... all of which improves your standing and reputation in the company and – even more importantly – the quality of your company's products.

The Market for Service Providers in the Aviation Context

The aviation market is mainly driven by the end users of its products and services, be they:

- individual customers (using air transportation to travel as passengers or to ship their goods, products, post and other items)
- governments (the military, coastguards, police, etc.)
- business or private organizations (such as rescue organizations and individual companies).

All of these will use aircraft to fulfil the roles allocated to them by their respective stakeholders.

Since the actual production of products and services in this market can more often than not be characterized as being very complex and costly, only those companies that thoroughly understand their relevant end users' needs, requirements and constraints are likely to survive in the aviation market.

Most service providers in the aviation context clearly depend on the overall development of air traffic that, in turn, depends on many factors such as the fuel price, the world's economic and overall political climate, just to name a few.

Figure 1.3 compares growth in both passenger traffic and available capacity across regions of the world in 2005. It is based on data taken from IATA (the International Air Transport Association) and only international scheduled air traffic is included in that data (domestic air traffic is excluded). Revenue Passenger Kilometres (RPK) measures actual passenger traffic, whereas Available Seat Kilometres (ASK) measures available passenger capacity.

The figure shows that, with more than 13 per cent, by far the strongest passenger growth rate could be observed in the Middle East region, whereas relative growth in Europe was lowest. Capacity growth was in line with traffic growth in North America and on average industry-wide. In the Middle East region relative capacity growth was significantly lower than traffic growth, while in Latin America the opposite was the case.

IATA expects the average annual passenger growth rate until 2008 to be highest on routes from Europe to the Middle East and within the Asia/Pacific region, with 7.7 per cent and 8.3 per cent respectively.

(April 2005 over April 2004)

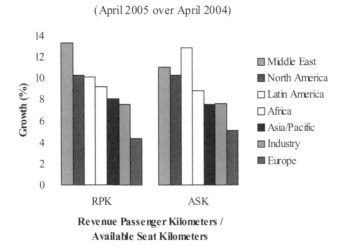

Figure 1.3   Passenger traffic and capacity growth

Preview of Operating Areas Covered

This book specifically covers services in the context of airline, airport and aircraft manufacturing operations. In the following, each of these three types of operations will be considered.

Aircraft Manufacturing Operations

Aircraft manufacturers Organizations such as Airbus, Boeing, Bombardier, Embraer, ATR, Saab, CASA, Eurocopter and many others develop, produce, support and maintain aircraft for all kinds of civilian and military customers.

Looking only at the civilian market, by far the biggest share of turnover is generated by dealing with airlines. Hence, airlines can be considered the most important direct customers of aircraft manufacturers overall.

Since developing an aircraft has a very long lead time and is a very complex and, therefore, costly matter, it is absolutely essential for manufacturers to keep in constant touch with their direct customers and the end users of their products and services (that is, passengers) in order to find out what will be their requirements and needs in the future, and to avoid investing in years of expensive development work only to come up with an aircraft that is not accepted by the market.

The outsourcing rate shown by leading manufacturers, both in terms of bought-in equipment/systems and engineering services, are clearly on the increase. In many cases this is already reflected by 60–80 per cent of the turnover from selling an airplane being consumed by bought-in products and services.

Suppliers and sub-contractors of aircraft manufacturers Suppliers that predominantly sell their products and attached services to aircraft manufacturers or intermediate system integrators are concerned to better meet their direct customers' technical and procedural requirements at a lower price than competitors. Companies with world reputation such as Honeywell, CAE, Thales, MTU, Rolls Royce, Liebherr or Alenia find themselves in a position where they can no longer dictate their terms and conditions to aircraft manufacturers; competition is fierce and suppliers usually have to apply their customers' rules in terms of the procurement process.

Sub-contractors that predominantly sell services in the form of engineering resources or consultancy to aircraft manufacturers (such as Seditec, Telelogic, Esterel, Aeroconseil, IBM and many others) find themselves not only in fierce competition, mainly based on price, but also they are pressed towards delivering lower value added services with very short-term contracts. This allows aircraft manufacturers to pay less for the services bought, remain highly flexible, secure their own personnel's jobs and fill high value added posts with internal staff. Sub-contractors, on the other hand, have to flexibly offer comprehensive and integrated solutions across all relevant customer sites if they want to prevail.

Airline Operations

Airlines 'Traditional' airlines (such as Lufthansa, Air France, British Airways, American Airlines, United, Air Canada, Quantas, Singapore Airlines, Emirates, All Nippon Airlines, Aero Mexico, Iberia, LAN Chile, Northwest Airlines, Alitalia and many others) are working hard to cope with the volatile market and get their service offerings right to meet their customers' needs, all this in a climate of tremendous price pressure in the lower margin segments with many 'low-cost' airlines successfully drawing away private and increasingly also business customers.

'Low-cost' airlines (such as Frontier, Blue, Ryanair, easyJet, Air Berlin, Deutsche BA (dba), Jetstar, JetBlue, Southwest Airlines, America West, Eurowings and others) have quite successfully focused on crushing operating costs by:

- dramatically reducing aircraft turnaround times (and therefore being able to squeeze more flights into one day)
- using smaller airports (which charge them lower fees)
- compressing the booking and check-in processes (Internet booking only, no tickets but just booking numbers, no seat reservations)
- reducing the range of inflight services offered to passengers.

On the other hand, low-cost airlines have opened new markets for people who were hitherto denied flying because they simply could not afford to pay traditional airlines' ticket prices. Furthermore, an ever-increasing number of smaller airports are now being served that may be situated closer to the real destinations of customers. What makes travelling 'low-cost' even cheaper is that passengers can usually park for free or at least for only small fees at the smaller airports used by low-cost carriers. One tremendous example that low-cost services are viable in the long term is Southwest Airlines, which has been at the low-cost forefront for 38 years and is now one of the biggest airlines worldwide in terms of passengers served.

The common belief that low-cost carriers could not financially survive in the long term because they do not charge their passengers enough money to be able to finance their operations is still widespread. Just to show that this does not always hold true, consider the following example. According to the US Department of Transportation's *Fourth Quarter 2004 Airline Financial Data* report, US Airways and Northwest (who can be considered traditional airlines) showed quarterly operating losses of $142.6m and $221.9m respectively. In contrast, during the same period, the low-cost carriers Southwest and JetBlue showed quarterly operating profits of $119.1m and $12.2m respectively.

'Specialized' airlines like PrivatAir have been seizing the opportunity of major companies needing flight services between their various sites worldwide. Such companies face the choice of either running their own flight operations (usually very expensive), using existing traditional or low-cost airlines, or outsourcing that service to an airline specialized in customized business shuttle flights. The latter may well offer tremendous timesaving, higher flexibility and lower costs depending on the circumstances.

'Cargo' carriers represent a rapidly increasing segment of the market. Many traditional airlines run their own cargo operations (such as Lufthansa and Singapore Airlines), either directly or as spin-offs. However, it is above all private postal service companies such as UPS and FedEx that play an increasingly important role worldwide, creating their own regional hubs with tremendous business impulses for local service providers and the job market in general. In Europe, for instance, Brussels (Belgium) and Leipzig (Germany) are examples of such regional hubs.

Suppliers and sub-contractors of airlines   Such organizations may be any one of the following:

- food preparation and delivery companies
- hotel chains
- crew shuttle services between the airport and those hotels that are consistently used by specific airlines for their crews (at lower charges)
- technical maintenance and repair services (for the aircraft an airline operates)
- aircraft leasing companies that provide an increasing number of airlines with individual fleet solutions.

Airport Operations

Airport services   By far, most larger airports are clearly run as profit organizations, with local towns, communities or countries usually holding an interest in those privatized operations. One example is the company Fraport AG which successfully runs the Frankfurt am Main Airport in Germany, one of the biggest and busiest European airports. Because there are many customers of airport services, such as airlines, shops, travel agencies and many other service providers, it is worthwhile to run airport services and provide these companies with the infrastructure they need to run their own operations.

In order to give an example of airport growth in terms of numbers of flights and passengers, Figure 1.4 compares four major German airports. Frankfurt and Munich can be considered as being mainly used by traditional airlines. Hahn and Luebeck serve as low-cost airports and were converted from a military air base and a general aviation airfield respectively. Although Frankfurt is much bigger in terms of passengers, Munich is the leading German airport in domestic business travel. Both airports showed significant growth rates in 2004, both in terms of passengers and flights. It does not come as a surprise that Hahn and Luebeck are much smaller. It seems that a large part of low-cost traffic growth in Europe is achieved by low-cost carriers continuously extending their network, adding new and hitherto unused small airports, as well as investing in a small number of 'low-cost hubs' such as London Stansted.

Considering the average number of passengers per flight across all four airports, Frankfurt served 107 passengers per flight, Munich 75 passengers, Hahn 112 and Luebeck 143. This can be argued as being in line with the four airports' profiles: Frankfurt is used for both domestic and international (also to a high percentage intercontinental) flights, whereas Munich shows a higher ratio of domestic business flights to international flights, so that smaller aircraft will be operated. Hahn and

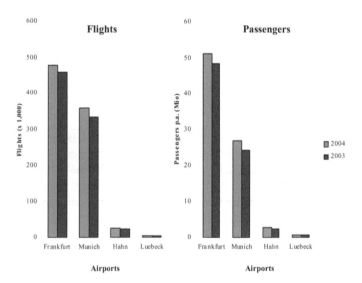

Figure 1.4   Comparison of major German airports

Luebeck reflect the fact that by far the most flights are operated by Ryanair using Boeing 737s, usually with a high load factor (very few empty seats). In the case of Hahn, Ryanair currently offers direct flights to 27 European destinations.

Suppliers and sub-contractors of airport services   These are, for instance, security companies such as Securitas, cleaning, building and construction companies, as well as many other service providers that are needed to run and maintain the highly complex infrastructure an airport represents.

How to Use This Book

The book contains three main parts. First, a small number of *proven and essential theories and frameworks* are presented that are internationally accepted as state-of-the-art in the fields of services marketing, management of operations and change management. This part offers additional information about the underlying theories and frameworks for the interested reader (Chapter 2).

Second, a *practical step-by-step method* is described, the 'service quality cycle', to show how exactly any service provider can implement – with little effort – an effective system to measure what matters most to his/her customers and how he/she is perceived to perform along those valued standards over time. This part is intended to give concrete suggestions to the reader as to how such a system can be implemented and how concrete actions can be derived from the results of such measurements in order to manage service quality as perceived by customers (Chapter 3).

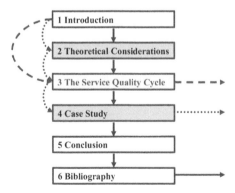

Figure 1.5   How to use this book

Third, a *recent case study* taken from an internal service provider within a European aircraft manufacturer is presented and critically evaluated. This case study offers useful insights in an example implementation of the method described in the second part of the book, culminating in the lessons learned from this practical experience (Chapter 4).

The book consists of self-contained chapters, which allows you to select the appropriate parts of the book only. Three different ways of going through this book are suggested, depending on what are your expectations from the book (see Figure 1.5):

1. You are in a hurry and want to urgently control/improve your service quality? Go directly to Chapter 3 and apply the method suggested there as appropriate to your individual situation.
2. You want to control your service quality but wish to gain more insight into the underlying theories/frameworks of the proposed method and the resulting practical experience? Read Chapter 2 first to learn more about the underlying theories/frameworks, and then proceed to the method proposed in Chapter 3. Finally, read about practical experience with the proposed method in a recent case study from an aircraft manufacturer.
3. You are interested in the fields of applied services marketing and/or service quality management in general? Read through the entire book in the order presented).

In order to facilitate implementing the suggested method, the appendix at the end of the book offers a summary of all checklists from Chapter 3 and can be used as an implementation checklist that can be easily adjusted to individual needs and circumstances.

Chapter 2

Theoretical Considerations

Chapter Summary

After the first chapter's brief overview of the content, context and usefulness of this book, the purpose of this chapter is to give the reader deeper insight into some of the state-of-the-art theories and frameworks that were selected to form the bases of the method suggested in Chapter 3. Although not all of these theories are directly incorporated in the proposed method as such, it is argued that it is very useful to include them in the present theoretical considerations and review them from an aviation perspective to gain a better understanding of relevant areas of service quality management.

In order to do this, the chapter first covers some general service issues that underline specific properties of service operations and their marketing. Then a comprehensive framework as to how superior service quality can be systematically created is presented, followed by the concept of performance objectives, their effects, measurement and visualization over time. Also, some change management issues are discussed, since every major change needs to be managed in some form to minimize harmful frictions, conflicts and resistance. Finally, the chapter covers specific quality issues in the field of aviation, including some relevant standards, prizes, awards, rankings and reports.

General Service Issues

This section covers some general service issues that underline specific properties of service operations and their marketing, since those operations differ significantly from operations that predominantly produce goods or products, mainly because of the inherent properties of services.

In some industries in the aviation context, for instance in the airline industry, emphasis on service quality is not a new idea. However, in other areas such as internal process, method and tool support for designer teams within an aircraft manufacturer (subject of the case study described in Chapter 4) it may well be considered an innovation or at least may not be systematically dealt with. There are still many fields in the aviation context where the focus is still very limited to the delivery of specified products or services, and service quality is not at all managed in an effective manner but rather depends on the performance of individual customer contact employees.

Definitions and Properties of Services

Palmer (1994) defines any *service* as 'the production of an essentially intangible benefit, either in its own right or as a significant element of a tangible product, which through some form of exchange satisfies an identified consumer need'. He identifies the following pure service features:

- *intangibility* (services cannot be touched)
- *inseparability* (production and consumption take place at the same time)
- *variability* (of both process and outcome)
- *perishability* (services cannot be stored for later consumption)
- *no ownership* (services cannot be possessed; no ownership is transferred due to their intangibility and perishability).

Santesmases (1994) identifies similar service features and argues that some of the biggest problems service providers are facing are the fact that services cannot be stored (even though irregular demand must be met) and that it is very hard to estimate the worth of services. Therefore, he argues that pricing and financial accounting of services is difficult.

De Vicuña Ancín (1999) makes the distinction between:

- services that are *related to a tangible product* (for example, distribution, financing and technical assistance)
- services that are *annexes to a tangible product* (such as maintenance and other after-sales services)
- services *not related to any tangible product*.

Typical service features, he argues, affect a company's *distribution strategy* (direct sale is the most common form), its *pricing strategy* (especially with less standardized services where customers need to be educated and given criteria of evaluation), its *sales strategy* (there is a trend to customize more), its *product strategy* and its *promotional strategy* (geared towards confidence-building).

Engelhardt (1996) sees *optimal service* in an area of conflict between customer satisfaction and service efficiency (see Figure 2.1). If the service is very useful to the customer and very cheap, he argues, the customer will be satisfied but the company does not make profit so that the service is not sustainable in the long run. If the service is very lucrative for the company but the customer thinks the price paid for the service is too high, the service will not be sustainable either. Thus, service offers should be positioned in the area of the service optimum, at least in the medium and long run.

Slack, Chambers and Johnston (2004) define all *operations* as 'a set of business processes which often cut across functionally based micro operations'. They argue that most operations produce some mix of tangible products and less tangible services, and only few operations produce either products or services alone. They

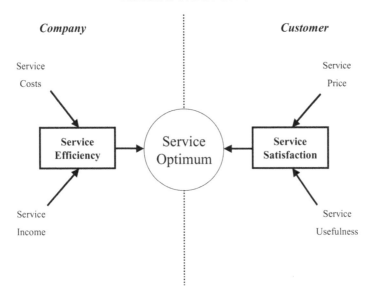

Figure 2.1   The service optimum
Adapted from Engelhardt (1996)

further claim that 'all operations, no matter how well they are managed, are capable of improvement'. However, 'all operations ... need some form of performance measurement as a prerequisite for improvement'.

In the frame of this book, 'service operations' are defined as operations that predominantly produce less tangible services. This definition covers both *external services* that are delivered to external customers outside the own organization and *internal services* that are delivered to internal customers inside the own organization.

Service Positioning

According to Zeithaml and Bitner (2002) a service offering's *position* is 'the way it is perceived by customers, particularly in relation to competing offerings'. They argue that 'service positioning is useful in establishing a new service image, as well as for maintaining and repositioning existing services'. Also, the service design and specifications developed in the processes described above are 'inextricably intertwined' with the service positioning, insofar as design and specifications will affect the image customers have got about the service and, in turn, the positioning strategy will affect the way the service on offer is designed if the positioning is to be effective in the medium and long run.

Let us look at an example. Niki is the Austrian carrier established in November 2003 by Formula One motor racing driver Niki Lauda. The company is based in

Salzburg and Vienna and operates a mixed fleet of Airbus aircraft from the A320 family. Although the company can be argued to be a low-cost airline, it is proactively positioned not to be a traditional low-cost carrier but rather a 'high service low-fare' airline. On-board drinks and snacks are not only offered but they are free. So are newspapers and magazines.

Services can be positioned on a variety of dimensions, depending on the needs they address, their benefits, individual service features, their users, as well as how and when they are consumed. The crucial thing to be kept in mind is that services should be positioned on one or several dimensions that are *valued by customers*, that can serve to *distinguish* the service from competitors' offerings, and that can be delivered to customers *consistently*.

Zeithaml and Bitner (2002) propose, as an example, two possible strategies companies could follow, the first being positioning on the 'five dimensions of service quality'. In this case, a company would focus on one or more of:

- reliability
- responsiveness
- assurance
- empathy
- tangibles.

It will thus position its service offering accordingly; for example, as being more reliable and more responsive than competitors' offerings.

Alternatively, a company may choose to concentrate on one or more of the 'dimensions of the service evidence':

- *people*: service contact employees and other customers
- *physical evidence*: tangible communication, price, physical environment, guarantees
- *process*: flows of activities, steps in the process, flexibility of process.

A service could be positioned, for instance, as being more effective and flexible as competing services and carried out by friendlier contact employees than other service providers have on their payroll.

The Expanded Marketing Mix for Services

Compared to the traditional marketing mix (or 4 Ps – product, place (distribution), promotion and price) for products, the marketing mix for services has to be expanded to account for additional complexity due to the specific features and characteristics of services.

Zeithaml and Bitner (2002) argue that service customers, because services are intangible, will 'often be looking for any tangible cue to help them understand the nature of the service experience'. Therefore, they suggest the additional elements

– people, physical evidence and process (mentioned above) – have to be included in marketing mix considerations (see Table 2.1).

Table 2.1   Expanded marketing mix for services

Product	Physical good features
	Quality level
	Accessories
	Packaging
	Warranties
	Product lines
	Branding
Place	Channel type
	Exposure
	Intermediaries
	Outlet locations
	Transportation
	Storage
	Managing channels
Promotion	Promotion blend
	Salespeople
	– Number
	– Selection
	– Training
	– Incentives
	Advertising
	– Targets
	– Media types
	– Types of advertisements
	– Copy thrust
	Sales promotion
	Publicity
Price	Flexibility
	Price level
	Terms
	Differentiation
	Discounts
	Allowances
People	Employees
	– Recruiting
	– Training
	– Motivation
	– Rewards
	– Teamwork
	Customers
	– Education

	– Training
	Communicating culture and values
	Employee research
Physical evidence	Facility design
	– Aesthetics
	– Functionality
	– Ambient conditions
	Equipment
	Signage
	Employee dress
	Other tangibles
	– Reports
	– Business cards
	– Statements
	– Guarantees
Process	Flow of activities
	– Standardized
	– Customized
	Number of steps
	– Simple
	– Complex
	Level of customer involvement

Adapted from Zeithaml and Bitner (2002)

Companies providing services have to carefully balance and optimize their specific services' marketing mix. Elements of the marketing mix must be consistent with other elements and real-life capacities or constraints. For example, if a company promises (promotion) high value, best-in-class service (product) at the lowest price (price) with incredible warranty conditions (product), those communications are likely to result in customers' disbelief rather than higher usage rates of the service provided. Also, if a company advertises (promotion) cheap (price) and responsive service (process) by well trained and friendly staff (people), but the service customers are actually experiencing is not-quite-so-cheap with long queues and they are served by unfriendly staff who do not seem to know what exactly they are doing, customers will be disappointed and lose confidence in all communications by this company.

Both customer disbelief and disappointment will probably result in the customer searching for the services needed elsewhere. It is particularly important to build up customer confidence in providing services, as the latter are largely intangible.

The final marketing mix for market introduction must be agreed upon by all internal stakeholders to avoid harmful inconsistency between individual elements of the marketing mix or between the marketing mix, and real capacities and constraints of the company.

Having discussed some general issues to do with service properties, service operations and their marketing, the next section will introduce a comprehensive framework of how superior service quality can be systematically created.

Example: PrivatAir

PrivatAir is an important operator in the business aviation sector and, based in Geneva, operates an entire fleet of more than 50 modern aircraft that are based in Europe and the US. These range from a Beech King Air to an executive B757. The fleet also comprises four new A319 aircraft, two of which are operated on behalf of Airbus to run the aircraft manufacturer's daily shuttle services between its sites in Hamburg, Filton and Toulouse.

PrivatAir A319 (Photo: Airbus)

Also, many other multinational companies use PrivatAir services for daily and/ or weekly business flights between their different sites or facilities at the direct convenience of their customers' employees. Delivering those services, PrivatAir contributes to tremendous time and cost savings for their customer companies.

All PrivatAir flights are usually business class or first class only and passenger comfort is taken very seriously. Since recently, long-haul direct flights between European and US destinations are offered using a long-range version of the A319, thereby addressing a market segment between corporate jet and traditional airline standard business class travelling. This segment is now better served by PrivatAir mainly because the passenger capacity is greater than is the case with most corporate jets and a lot of time and money is saved (compared to traditional commercial flights) by using the closest airport to the destination.

In June 2005, PrivatAir added a new Global Express ultra-long-range jet to their fleet that was made available for private charter immediately. Seating a total of 12 passengers, this aircraft allows for long-range services up to 11,000 km (such as from Geneva to Los Angeles), in all the comfort and luxury that their demanding business customers expect.

Service quality issues are of utmost importance for PrivatAir in order to keep offering world-class customized solutions to their customer companies and to successfully defend and further develop their market segment.

The Gaps Model of Service Quality

In this section, the 'gaps model of service quality', as proposed by Zeithaml and Bitner (2002), will serve as a framework for considerations as to how superior service quality can be systematically created. This conceptual model (see Figure 2.2) is very comprehensive and identifies sources of potential direct dissatisfaction, where customers do not get what they expect from a service (the Customer Gap, or Gap 5), or sources of direct dissatisfaction that lead to that perceived gap, where the company does not deliver what the customer expects (Provider Gaps 1–4).

The process of closing the *Customer Gap* (the difference between what the customer is expecting and what the customer perceives he or she gets from a service) can be subdivided into four internal company gaps or *Provider Gaps*:

- Provider Gap 1: not knowing what customers expect
- Provider Gap 2: not selecting the right service designs and standards
- Provider Gap 3: not delivering to service standards
- Provider Gap 4: not matching performance to promises.

In the following, the Customer Gap and each of those Provider Gaps, with their underlying factors, are examined in general, and some possible ways to close each

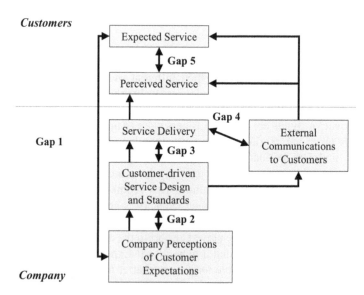

Figure 2.2   The gaps model of service quality
Adapted from Zeithaml and Bitner (2002)

individual gap are presented. Concrete examples from the aviation context are used to illustrate and review specific parts of this framework or related theories.

The Customer Gap

This is the difference between the service a customer expects, perhaps based on past experience, word-of-mouth communications or promises made by the company in one form or another, and the delivered service, as perceived by the individual customer (see Figure 2.3). Although it is argued as resulting to a large extent from the four Provider Gaps and, therefore, is called 'Gap 5', the Customer Gap will be considered first in order to better understand what this framework is all about – that is, reducing the crucial gap that is perceived by the customer and thus in one way or another influencing his future behaviour.

A passenger using an airline will have built up in their mind (mostly subconsciously) a set of expectations of the service likely to be encountered: these expectations will range along a number of criteria such as friendliness of staff, punctuality of the flight and cleanliness of the aircraft. These expectations may be mainly based on advertisements about the airline, word-of-mouth communications from other people who have already used the airline and perhaps the passenger's own past experience with the airline. During the actual delivery of the service (that is, during the journey) the customer will experience or perceive the service to be as

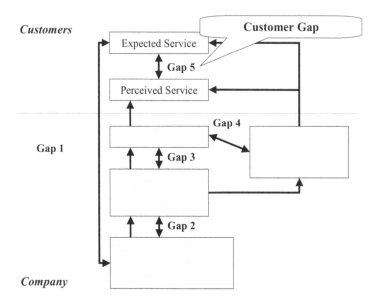

Figure 2.3   Closing the Customer Gap (Gap 5)
Adapted from Zeithaml and Bitner (2002)

expected or possibly quite different from what was expected (better or worse). The Customer Gap means that perceived difference.

Zeithaml, Berry and Parasuraman (1993) suggest the following model of the 'nature and determinants of customer expectations of service' to explain how customer expectations are developed (see Figure 2.4). They argue that the *expected service* lies somewhere between the *desired service* (the best service the customer hopes for) and the *adequate service* (the lowest level of service quality the customer would accept). A *zone of tolerance* separates those two service levels, where customers are neither delighted by the service nor feel disappointed about it but do not particularly notice service performance.

An individual customer's zone of tolerance increases or decreases depending on a number of factors, including factors controllable by the company, such as price. Customers' tolerance zones also vary for different service dimensions and there are differences between first-time service and recovery service situations. For example, if a company does not get the service right the first time, if a customer has complained about it, he or she will be even less tolerant the second time and the expected service level will have gone up in most cases. Also, a customer purchasing a low price service is likely to have lower service expectations concerning the quality of the service, than he or she would have if the price for the service had been higher.

A number of factors affecting the 'expected service' and its components have been identified. 'Desired service' is influenced by:

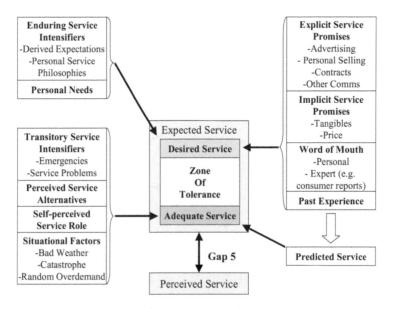

Figure 2.4   Nature and determinants of customer expectations
Adapted from Zeithaml, Berry and Parasuraman (1993)

- *enduring service intensifiers* – 'individual, stable factors that lead the customer to a heightened sensitivity to the service'
- *personal needs* (perhaps physical, social, psychological or functional)
- *explicit* and *implicit service promises* the company makes
- *word-of-mouth* communications by reference people, and the customer's own
- *past experience* with the service in question or similar services.

The latter three affect both the 'desired service' and the 'predicted service'. Hence, indirectly (via the predicted service), those factors influence the 'adequate service' level, too.

'Adequate service', on the other hand, is influenced by:

- *transitory service intensifiers* – temporary and usually short-term individual factors that make the customer aware of the need for a service
- any *perceived service alternatives*
- the *self-perceived service role* of the customer
- *situational factors*
- *predicted service.*

A number of ways have been proposed for services marketers to influence those factors, and thereby the service levels expected by the customer (see Table 2.2).

The second service level that is involved in creating the *Customer Gap* is the level of *perceived service* the customer experiences. This customer perception of service is influenced by the following factors (see Figure 2.5): service encounters, evidence of service, image and price.

Service encounters can be described as 'moments of truth' in so far as they represent, from the customer's point of view, the most vivid impressions of the service provided. On the other hand, this means that, from a company's viewpoint, each encounter 'represents an opportunity to prove its potential as a quality service provider and to increase customer loyalty'.

Service customers are searching for *evidence of service* in every interaction with the service company, because services are intangible. There are, as mentioned before (in Table 2.1) as the new elements of the marketing mix for services, three categories of service evidence:

- *people* (contact employees, the customer him/herself, other customers)
- *process* (operational flow of activities, steps in process, flexibility versus standardization, technological versus human)
- *physical evidence* (tangible communication, 'servicescape', guarantees and technology).

All three together contribute to 'tangibilize' the otherwise intangible service to the customer.

Table 2.2   Ways in which services marketers can influence factors

Controllable factors	Possible influence strategies
Explicit service promises	Make realistic and accurate promises that reflect the service actually delivered
	Ask contact people for feedback on the accuracy of promises made
	Avoid engaging in price or advertising wars
	Formalize service promises through a service guarantee
Implicit service promises	Assure that service tangibles accurately reflect the service provided
	Ensure that price premiums on important attributes are justified

Less controllable factors	Possible influence strategies
Enduring service intensifiers	Use market research to determine sources of derived service expectations and to profile personal service philosophies of customers
Personal needs	Educate customers on ways the service addresses their needs
Transitory service intensifiers	Increase service delivery during peak periods or emergencies
Perceived service alternatives	Be fully aware of competitor offerings and, where possible and appropriate, match them
Self-perceived service role	Educate customers to understand their roles and perform them better
Word-of-mouth communications	Identify influencers and opinion leaders and concentrate marketing efforts on them
Past experience	Use market research to profile customers' previous experiences with similar services
Situational factors	Use service guarantees to assure customers about service recovery, regardless of the situational factors that occur
Predicted service	Tell customers when service provision is higher than what can normally be expected so that predictions of future service encounters will not be inflated.

Adopted from Zeithaml and Bitner (2002)

The *image* of a service also influences customers' perceptions of the service. It is built up in customers' minds essentially through communications such as advertising, physical images or word of mouth, combined with actual experiences. The image of a service can and does usually affect customer perceptions of quality, value and satisfaction.

Last, the *price* of the service, too, is argued to influence the perception of service quality, value and satisfaction because services, due to their intangibility, are often difficult to judge before they are purchased, and price is frequently taken as a service quality indicator that influences expectations and perceptions among customers.

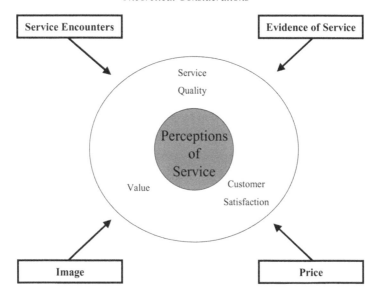

Figure 2.5   Factors influencing customer perceptions of service
Adapted from Zeithaml and Bitner (2002)

Because the four factors discussed above significantly influence perceived service, the following *strategies for influencing customer perceptions* have been recommended:

- Aim for customer satisfaction in every service encounter:
    - plan for effective recovery (see Günter and Huber 1996, below)
    - facilitate adaptability and flexibility
    - encourage spontaneity
    - manage the dimensions of quality at the encounter level.
- Manage the evidence of service proactively to reinforce perceptions.
- Communicate realistically and use customer experiences to reinforce images.
- Use price to enhance customer perceptions of quality and value.

Günter and Huber (1996) argue that *complaint management* must be conducted systematically and proactively (not just reactively), based on regular and non-superficial customer information, and geared towards both recovery of the individual customer's dissatisfaction, in the short term, and further development and improvement of the customer contact processes, in the medium and long term.

Otherwise, possible results are loss of trust among customers, the reporting of dissatisfaction to other customers by word-of-mouth communications, and, as a result, loss of customers to competitors. Therefore, they recommend:

Example: Airbus

Airbus, currently the world's leading aircraft manufacturer in the civilian market, has to manage a lot of complexity due to the geographical and transnational spread of its sites and facilities, mainly across Europe. Parts of the fuselage, the wings, stabilizers, vertical tail planes, etc., have to be transported by road, by ship and/ or by air to be further integrated step-by-step into an aircraft at one of the final assembly lines in Toulouse (France) or Hamburg (Germany).

Beluga transport (Photo: Airbus)

Less visible but just as important is the transnational integration of engineering and design work. Depending on the programme, development teams are spread across a large number of sites in several mainly European countries. This leads to many technical and organizational interfaces and internal customer supplier relationships between development teams, as well as centres of competence and excellence across Airbus.

Beluga landing (Photo: Airbus)

A380 transport (Photo: Airbus)

Such complexity can only be managed efficiently if there is a very high level of internal service quality between all relevant parts of the organization that fully takes into account internal customers' needs and requirements in a challenging transcultural environment within the company.

Hence, not only externally in relation to airlines but also internally between technical interface partners, service quality management issues are essential in order to produce better and safer aircraft in less time and for less money than competing aircraft manufacturers do.

Airbus A319 (Photo: Airbus)

- immediate recovery of the customer's problem (reactive repair function)
- improvement of the company's performance (proactive, longer-term)
- use of indicators of customer satisfaction as measurement for customer-oriented behaviour and a means of marketing control.

Provider Gap 1 – Not Knowing What Customers Expect

Returning to the gaps model of service quality, Provider Gap 1 is the difference between the service, as expected by the customer, and the company perceptions of what the customer expects (see Figure 2.6). If the company does not even know what customers expect, all its internal efforts to provide a service will not be directed at what really matters – customers' requirements.

An aircraft manufacturer needs to be very close to its customer airlines and know their markets, routes, types and locations of facilities used, existing aircraft, financial situation, strategic plans as well as many other things in order to identify 'aircraft solutions' (including service aspects such as training, maintenance and financial issues). Those aircraft solutions need to take into account the whole picture and strive to meet individual airlines' present and (even more importantly) future needs.

The reason why it is so important to meet future needs is that the development of new aircraft and related services has a long lead time and is therefore very expensive. Some aircraft manufacturers have ceased to exist because (amongst other things)

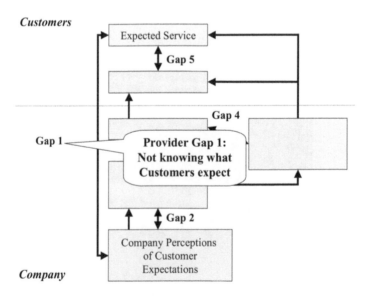

Figure 2.6   Closing Provider Gap 1
Adapted from Zeithaml and Bitner (2002)

they spent their resources developing aircraft that were not accepted by the market or because they failed to offer complete aircraft solutions needed by specific airlines.

The following key factors leading to Provider Gap 1 have been identified. They are related to *marketing research* and to *building customer relationships*: we now look at these in turn.

Key factors related to marketing research:
- insufficient marketing research
- inadequate use of marketing research
- lack of interaction between managers and customers
- insufficient communication between contact employees and managers.

Marketing research is not the only way for a company to explore its environments, but certainly a major one that should not be underestimated. Research should be done on an ongoing and systematic basis and, when needed, special studies should be carried out in areas of interest that are not sufficiently explored by ongoing market research.

There is strong supporting evidence that carefully conducted and directed market research helps in the processes of market segmentation, targeting the right segments and positioning a company's offerings within those segments (Kotler et al. 1996, Zeithaml and Bitner 2002, Dibb et al. 1997). The following are among the most promising areas:

- *complaint solicitation* – to identify and attend to dissatisfied customers, and to identify common service failure points (qualitative continuous research);
- *critical incident studies* – to identify 'best practices' at transaction level, customer requirements as inputs for quantitative service, common service failure points, and systemic strengths and weaknesses in customer-contact services (qualitative periodic research);
- *requirements research* – to identify customer requirements as input for quantitative research (qualitative periodic research);
- *relationship surveys* – to monitor and track service performance, assess overall company performance compared with that of competitors, determine links between satisfaction and behavioural intentions (quantitative annual research);
- *key client studies* – to create dialogue and close the loop with important customers (qualitative and quantitative annual research),
- *lost customer research* – to identify reasons for customer defections (qualitative continuous research);
- *future expectations research* – to forecast future expectations of customers, and develop and test new service ideas (qualitative and quantitative periodic research).

Key factors related to building customer relationships:
- lack of market segmentation
- focus on transactions rather than relationships
- focus on new customers rather than existing customers.

There is strong evidence that, even in areas where there are presumably enough customers about to build on many transactions rather than long-term relationships, the latter approach pays off in the medium to long run because – among other reasons – satisfied long-term customers spend more over time, try new products offered by the company and, very importantly, spread positive word-of-mouth communications for the company among potential future customers.

Relationship marketing (or relationship management) is a philosophy of doing business, a strategic orientation, that focuses on keeping and improving current customers, rather than on acquiring new customers. This philosophy assumes that consumers prefer to have an ongoing relationship with one organization than to switch continually among providers in their search for value services.

Based on this assumption and the fact that it is usually much cheaper to keep a current customer than to attract a new one, many successful marketers have proposed effective strategies for retaining customers that are beneficial for both sides involved.

Zeithaml and Bitner (2002) propose the following *retention strategies:*

- monitoring relationships;
- recovery (retaining customers when things go wrong):
  - track and anticipate recovery opportunities
  - take care of customer problems on the front lines
  - solve problems quickly
  - empower the front line to solve problems
  - learn from recovery experiences;
- customer appreciation.

Monitoring customer relationships includes using a well designed customer database, but avoiding to make the customer feel being spied upon. Recovery has to be planned for in order to be effective, and empowerment of front-line employees is a crucial issue here. Lastly, customers must feel appreciated and never receive the impression from front-line employees of the company that they are just a customer who is 'lucky' to get the company's 'help'.

One typical example, where all three strategies could be followed for the benefit of the service provider (and its customers alike), would be an airline that has to cancel a flight for technical reasons without an immediate alternative flight it could offer to passengers. The names of the passengers concerned should be marked as having gone through this disappointing situation in the airline's customer database so that efforts to turn around angry customers later on can be directed to them. Perhaps a voucher of some sort sent by post one day later, along with an apology letter, could do the trick.

The latter step is already part of the recovery measures. To be able to react on the spot and limit customer frustration, the airline must have some sort of effective recovery procedure for such cases in place and front-line employees sufficiently empowered to carry out these plans whenever needed. Such 'emergency plans' ought to be established by a small group of experienced key employees who know from past experience the potential situations that will need immediate recovery and what will be the best practice approaches to deal with them.

Front-line employees would then need to be trained and empowered to handle such crises appropriately. Each recovery experience should be debriefed after the situation is solved and the results recorded in some form so that this experience can contribute to continuous improvement.

Finally, employees must show customers honest respect and appreciation for who they are as persons and consumers of the services offered. This helps create the right mindset to deliver services and is a prerequisite (not a guarantee though) for a healthy and potentially long-term customer relationship. This is not to say, however, that cabin staff have to take any kind of bad behaviour from (say) drunken passengers or cannot (politely) refuse unrealistic demands made by customers.

Provider Gap 2 – Not Selecting the Right Service Designs and Standards

Provider Gap 2 is caused by inappropriate translation of the company's perception of what the customer expects into a company's internal service designs and standards (see Figure 2.7). Not everything about services can and should be standardized (for example, flexibility and front-line empowerment is necessary in many services) but certain routine activities need standardization to achieve a certain consistency of service and to be able to benefit from synergies.

Airlines using airport facilities have very clear expectations from the airport's ground support teams. Their main interests may be to turn an aircraft around safely, as quickly as possible and for the lowest price. Airport operations not only need to be aware of these expectations from customer airlines but they need to translate them into corresponding internal service designs and standards. In other words, they need to find out what are the best practices to safely yet speedily turn around an aircraft (including re-fuelling, safety checking, luggage handling, etc.) and formulate corresponding standard procedures that will be applied consistently by all their ground support teams.

The following key factors that lead to Provider Gap 2 have been proposed. They are related to *service standards* and *service leadership*.

Key factors related to service standards:
- inadequate standardization of service behaviours and action
- absence of formal process for setting service quality goals
- lack of customer-defined standards.

Example: Fraport AG

Frankfurt am Main airport in Germany is one of the busiest European airports, with over 50 million passengers carried each year.

Frankfurt Airport MD-87 (Photo: Fraport AG)

The airport has created about 65,000 direct and 120,000 indirect jobs in the region. It is very successfully run by the Fraport AG. However, night-flight restrictions and relatively high airport fees makes it somewhat less attractive to cargo service providers and low-cost airlines.

The Fraport AG also holds about 70 per cent ownership of the Frankfurt/Hahn Airport, which is in the countryside about a one hour's drive west of Frankfurt (not at all close to the city). This airport has been very attractive because it can stay open 24 hours a day and charges very low fees. Therefore, it has seen tremendous increase in activities with both cargo services and low-cost airlines, above all Ryanair.

Frankfurt Airport rail and road connection (Photo: Fraport AG)

In June 2005, Lufthansa signed a long-term contract over a duration of 65 years to use a 42,000m² area of the Frankfurt am Main Airport so that they can build a maintenance hall that will be large enough and fitted to accommodate four A380s or six B747s simultaneously for maintenance purposes. The building is scheduled to be finished just in time for the arrival of Lufthansa's first two A380s in autumn 2007.

Also, like in Munich, Lufthansa has opened a brand-new 1,800 m² (19,375 sq ft) first class terminal in Frankfurt that offers a wide range of services to their first class passengers. These services include the provision of furnished luxurious rooms with spacious bathrooms, a wide range of movies and video games, fully equipped offices, massage facilities, hot meals made to order and chauffeur-driven transport (Mercedes/Porsche) directly from the new terminal to the aircraft.

Fraport ground support (Photo: Fraport AG)

Both external and internal service quality is a key issue for the airport's operations to be run effectively and efficiently both within the Fraport AG and across all companies that are involved directly and indirectly in their airport operations.

It has been suggested that applying a systematic process for setting customer-defined standards helps to close Provider Gap 2 and, hence, contributes to high service quality (see Figure 2.8).

The first step is to explore and identify what is the sequence of service encounters as desired by the customer. Then, those customer expectations have to be translated into concrete actions or behaviours. This step is greatly enhanced if customers are involved ('partnering'). Next, behaviours and actions for standards have to be selected. The most essential criteria for the creation of standards are:

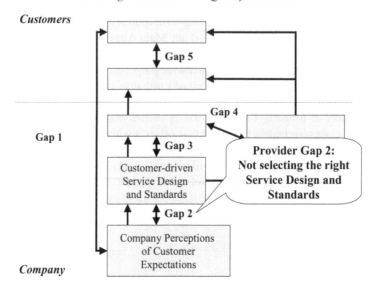

Figure 2.7   Closing Provider Gap 2
Adapted from Zeithaml and Bitner (2002)

- that they are based on actions and/or behaviours that are important to the customer
- that the standards cover performance that needs to be improved or maintained
- that the standards cover behaviours and actions that employees can improve
- the standards must be accepted by employees
- they should be predictive rather than reactive (for example, they should not be based solely on complaints)
- they ought to be challenging but realistic.

Then it should be decided which standards should be formulated as hard standards (quantitatively measurable, such as complaint handling time) or soft standards (qualitatively measurable, such as recovery satisfaction). Both *soft* or *qualitative standards* in the field of customer satisfaction in different situations and *hard* or *quantitative standards* have to be selected, the latter concerning delivery in time at the right quality, sticking to the agreed timetable and the negotiated specifications. It then follows that feedback mechanisms have to be developed to regularly measure and review the standards selected. It is critical that the company adopts a customer perspective, as opposed to a company perspective, to develop such feedback mechanisms.

The next step involves establishing measures and target levels. This step is important because the company needs to compare actual measurement results

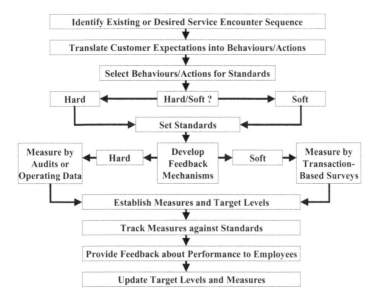

Figure 2.8   Process for setting customer-defined standards
Adapted from Zeithaml and Bitner (2002)

against challenging but realistic measures to see how it is doing and whether targets have been achieved.

The following step is that the results should be communicated to the corresponding employees and it should be discussed whether performance was worse or better than expected, or, alternatively, whether or not the standards set were unrealistic and need updating.

Hope and Mühlemann (1997) claim that productivity measurement is crucial to the level of service quality delivered because it means feedback to direct improvement efforts. There are certain areas which will be of special interest to any company:

- *efficiency* (the 'manner in which inputs are used to produce outputs')
- *effectiveness* (the extent to which the objectives are achieved)
- *utilization* (the 'percentage of time that resources are employed')
- *productivity* (the 'ratio between inputs and outputs').

Kleinaltenkamp, Fleiß and Jakob (1996) argue that 'only companies that are really customer oriented will survive in the long term … but most companies have still a long way to go to meet this requirement, especially among companies in the business-to-business sector'. They suggest a model of the basic principle of 'customer integration' (see Figure 2.9). Their model basically shows that if a customer has a problem, then this problem can best be solved by customer integration, meaning the integration of the customer in the process of problem-solving, as opposed to the

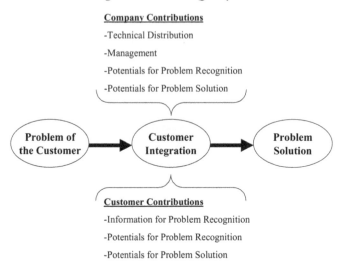

Company Contributions

-Technical Distribution

-Management

-Potentials for Problem Recognition

-Potentials for Problem Solution

Problem of the Customer → Customer Integration → Problem Solution

Customer Contributions

-Information for Problem Recognition

-Potentials for Problem Recognition

-Potentials for Problem Solution

Figure 2.9   Basic principle of customer integration
Adapted from Kleinaltenkamp, Fließ and Jakob (1996)

company trying to solve the given problem independently from the customer. They argue that the customer is a valuable resource in the process insofar as he holds relevant information for problem recognition, as well as potential for both problem recognition and solution. The serving company, in turn, contributes with its technical distribution capacity, management know-how, and employees' own potential for problem recognition and solution.

By blending all the information and potentials of both the company and the customer, both problem recognition and solution are significantly enhanced. This integration is also likely to have positive effects on long-term customer retention because it helps the company to understand its customers better and to develop closer long-term relations.

The second set of key factors leading to provider gap 2 is related to service leadership.

Key factors related to service leadership:
• inadequate service leadership
• lack of recognition that quality service is a profit strategy
• imbalanced performance scorecard.

The third point refers to having set the right mix of standards and measurements. As suggested above, both hard and soft standards must be defined that are easy to measure and, above all, matter to the customer.

Example: easyJet

In 2002, easyJet, who started operations in 1995 as the first British low-cost provider, surprised the aviation community with their order of 120 Airbus A319s (plus 120 options) to replace their ageing fleet of Boeing 737s. In the meantime, the first new A319s have completed their first year of service with the dynamic low-cost airline. As repeatedly stated by company representatives, the A319s have shown unexpected levels of reliability from their first flight hours on, far better than was the case with their existing fleet. Ray Webster, easyJet Chief Executive, stated that the introduction of the new Airbus aircraft had been the smoothest and most trouble-free in his 40 years in the business. The turnaround time – one key indicator of productivity, especially for low-cost operators – could be reduced to less than 20 minutes (!) with the new aircraft.

easyJet A319 (Photo: Airbus)

In Europe, the airline's main competitor is Ryanair. In France, for instance, easyJet is leading the race in terms of passengers carried. In Germany, Ryanair is leading the way. Both airlines have clearly benefited from the battle between Boeing and Airbus to increase their market share in the low-cost sector. For Airbus, easyJet was the first major low-cost customer so that the airline was offered tremendous price reductions and a very high degree of technical customization of the aircraft when buying their new fleet of A319s. For instance, the number of seats could be increased by meeting the safety requirement of having a second emergency exit door on each side.

easyJet has apparently sold on at least some of its new A319 aircraft to a leasing company from which it leases them back, thereby increasing its liquidity and making a clear profit from the Airbus deal. It has been estimated that the price per aircraft was about $22m and that they were sold on for $26m–$27m, at a real market value of $30m.

Like other low-cost airlines, easyJet is facing fierce competition and must be fully aware of what customers want and need, and make serious efforts to meet their requirements better than competitors do. Hence, service quality management over time is one major concern to the airline.

Provider Gap 3 – Not Delivering to Service Standards

Provider Gap 3 is caused by not appropriately applying the company's internal service designs and standards to actual service delivery (see Figure 2.10), even if the former are based on the right perceptions of real customer expectations.

If an airline has made a lot of effort to find out the needs and wants of their different types of customers and has also translated those into its service design and standards, there is one step missing to close the cycle: those standards must be applied consistently. By definition, if those standards reflect what customers expect, any services that do not meet them fail to meet the identified customer expectations. This potentially leads to lower customer satisfaction levels than would be possible.

The following key factors that lead to Provider Gap 3 have been identified. They are related to *employees, intermediaries, customers, demand and capacity*, as well as *international marketing*.

Key factors related to employees:
- ineffective recruitment
- role ambiguity and role conflict
- poor employee-technology-job fit
- inappropriate evaluation and compensation systems
- lack of empowerment and teamwork.

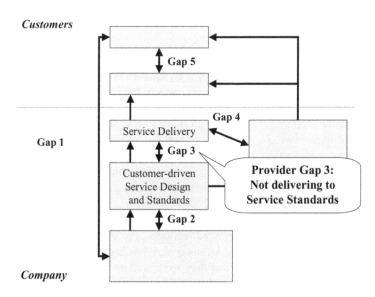

Figure 2.10   Closing Provider Gap 3
Adapted from Zeithaml and Bitner (2002)

The above factors can be attributed to the way a company appreciates and deals with its human resources. It can be argued that many companies have high employee turnover rates because often people are employed who are not sufficiently qualified for a job and who do not fit in the company for other reasons. Also, once an employee is hired, he or she is often not given the right opportunities for training nor the needed tools and maybe guidelines.

Finally, if the reward system in use does not reward what is needed (such as high service quality, taking responsibility, teamwork and process thinking) but only what is easy to measure (such as hours sat behind the desk), then employees are likely to become frustrated and they may feel the urge to leave the company.

The following *human resource strategies* have been proposed to ensure that the above issues are addressed and that a positive service culture is fostered amongst employees:

- *Hire the right people* – by competing for the best people, hiring for service competencies and service inclination, and being the preferred employer.
- *Develop people to deliver service quality* – by training for technical and interactive skills, empowering employees, and promoting teamwork.
- *Provide the needed support systems* – by developing service-oriented internal processes, providing supportive technology and equipment, and measuring internal service quality.
- *Retain the best people* – by including employees in the company's vision, treating employees as customers, and measuring and rewarding strong service performers.

Key factors related to intermediaries:
- channel conflict over objectives and performance
- channel conflict over costs and rewards
- tension between empowerment and control
- channel ambiguity.

Key factors related to customers:
- customers lack understanding of their roles
- customers are unwilling or unable to perform their roles
- customers are not rewarded for good performance
- other customers interfere
- market segments are incompatible.

Key factors related to demand and capacity:
- failure to smooth the peaks and valleys of demand
- overuse of capacity
- relying too much on price to smooth demand.

Since services in general cannot be stored and customers 'participate' in the actual service delivery, there will be situations where capacity of a company to deliver a service does not equal demand for that service by customers. In fact, this is hardly ever the case. Three *strategies for matching capacity and demand* have been suggested:

- shift demand (know how to move service demand away from peak periods)
- flex capacity (know how to adjust the number of employees to service demand)
- waiting-line strategies (know how to motivate customers to wait their turn).

An airline, for instance, may try to shift demand by offering cheaper flights at times when demand is expected to be low. However, in cases where demand is not very easy to shift (such as for business flights) the airline may try to flex capacity by using different aircraft (and cabin crews) in the short term and adjust the flight schedule to offer more flights in the morning and evening so that it can cope with the higher levels of business customers' demand. Finally, if customers have to wait, the airline may try to make this waiting time as pleasant as possible, perhaps by inviting customers who are waiting for a delayed flight into the airline's lounge or offering vouchers for drinks.

Key factors related to international marketing:
- legal barriers
- cultural barriers and differences.

Provider Gap 4 – Not Matching Performance to Promises

Provider Gap 4 is caused by differences or inconsistencies between what has been promised by the company in one form or another, and what is actually delivered to the customer (see Figure 2.11). In other words, performance is not matched to promises.

Since companies such as airlines seek to gain market share and attract customers, it is in their interest to communicate to the market that they know what customers want and that they have aligned their services to those expectations. However, in the longer run this only works if the airline actually delivers what is promised. The reason for this is that external communications directed at the customer are one of the most important factors that influence the level of expected service. The more promising the advertisements, the higher the expectations and the greater the disappointment if the actual (perceived) service delivery fails to come up with those promises.

External communications also influence the level of perceived service because most customers – due to the intangible nature of services – subconsciously use those to 'tangibilize' the services; that is, to make them 'touchable'.

The following key factors that lead to Provider Gap 4 have been identified. They are related to *communications*, *pricing*, and *physical evidence*.

Example: dba

Deutsche British Airways (dba) was intended to gain British Airways' market share from Lufthansa on domestic German flights and had a history of 11 years of losses when it was surprisingly bought by the German entrepreneur Hans Rudolf Wöhrl in 2003. By drastically reducing leasing payments for their fleet of 737s, doing more efficient flight planning, focusing on business travellers and offering maintenance services for other carriers, the airline managed a spectacular turnaround. By the end of the business year 2004–2005, dba had achieved very positive operational results – sales of about €265m and (for the first time) profits of more than €1m. Currently, dba offers flights to eight German and two other European destinations.

dba 737s (Photo: dba)

One increasingly important route served by dba is that from Berlin to Bonn and back. This route is very particular, because it is mainly or at least very often used by German politicians and their family members who commute or travel between Bonn, the former capital of West Germany, and Berlin, the traditional and new German capital, with some government offices and departments still being split between both cities. Following the SKYTRAX Airline of the Year 2004 Survey, dba has been rated a three-star airline and hence can be considered to be amongst the best European low-cost carriers in terms of service/product quality.

In light of the fact that an important target segment for dba is that of German business travellers, and the fierce competition by other low-cost airlines serving some of the same routes, service quality management issues are very essential for dba in order to meet customers' needs and expectations better than competitors do.

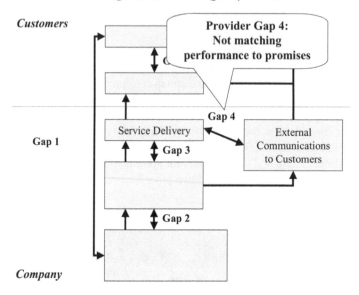

Figure 2.11   Closing Provider Gap 4
Adapted from Zeithaml and Bitner (2002)

Key factors related to communication:
- inadequate management of service promises
- over-promising in advertising and personal selling
- insufficient customer education
- inadequate horizontal communication
- differences in policies across branches or units.

Key factors related to pricing:
- assuming that customers hold reference prices for services
- narrowly defining price as monetary cost
- signalling the wrong quality level with an inappropriate price
- not understanding customers' value definitions
- not matching price strategy to customers' value definitions.

Key factors related to physical evidence:
- incompatible or inconsistent physical evidence
- over-promising through physical evidence
- lack of physical evidence strategy.

It is very important to develop effective physical evidence strategies, but relevant information from customer feedback is needed, such as how physical evidence affects customers and what their impressions are.

The following *approaches for matching service delivery and promises* have been proposed (see Figure 2.12). To better illustrate these approaches, an IT service provider within an aircraft manufacturer is considered as an example.

Managing service promises:
- make realistic promises
- offer service guarantees
- keep customers informed about changes
- negotiate unrealistic expectations
- set prices to match quality levels.

An internal IT service provider within an aircraft manufacturer that is about to install new IT infrastructure for a number of design teams would have to explicitly inform their customers about hardware costs, the performance and limitations of the new infrastructure, as well as their own regulations they need to apply when delivering the services in question. The latter may be crucial because in order to prevent loss of data and maintain full data integrity the aircraft manufacturer will have very strict rules, such as that new software or updates are not to be installed unless they have been thoroughly tested. This, in turn, will take time and customers must be made aware of this so that they do not expect the impossible and will not be disappointed later on. Service offer guarantees in terms of maximum waiting time for specific

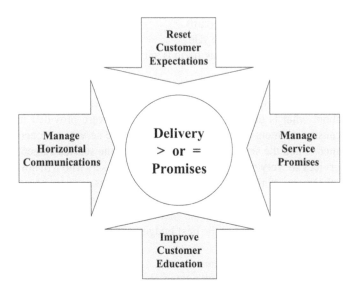

Figure 2.12   Approaches for matching service delivery and promises
Adapted from Zeithaml and Bitner (2002)

activities and/or minimum downtimes of the infrastructure and/or timeframe of availability of support personnel must be clearly stated and are usually part of a service agreement letter between the IT service provider and each customer team.

Resetting customer expectations:
- offer choices
- create tiered-value offerings
- communicate criteria for service effectiveness
- communicate realities in the service industry.

As mentioned above, customers of the IT service provider must be made aware of restrictions, regulations and constraints under which the provider has to operate. In the service offer made to customers, choices or options should be individually stated so that customers can see which services would be included at a certain price and which would not. If then a customer team only opts for service A and not B, that team is more likely to have reset its expectations to entitlement only to receive service A, because this is explicitly what they are paying for.

Clear criteria for service effectiveness should be negotiated with the customer teams. These criteria should be easy to measure and meaningful; that is, they should reflect real customer requirements.

Improving customer education:
- prepare customers for the service process
- confirm performance to standards
- clarify expectations after the sale
- teach customers to avoid peak periods and seek slow periods.

By means of personal contact with direct users of the services and by formal memos to customer teams' management the IT service provider should keep customers informed about what is going on and which services were provided in a specific period, because customers must be kept aware that the services are actually being used by someone and the service provider is fulfilling the service agreement letter.

Since services are very often delivered to individual users who will not necessarily talk about their use of the service (because sometimes it may be considered a matter of personal pride not to need the help of the hotline), it is easy for customer management to get the false impression that nobody really needs the services and that those costs could be saved in the next budget period.

Managing horizontal communications:
- align back office personnel with external customers
- open channels between advertising and operation
- open channels between sales and operations
- create cross-functional teams.

This issue is most relevant to larger service organizations. In the given example of an IT service provider within an aircraft manufacturer, it is more likely that the service provider itself is not such a large organization and most employees are likely to be in direct customer contact.

Also, advertising, sales and operations will (more often than not) be taken care of by the same IT managers. However, there is another challenge due to the transnational spread of most aircraft manufacturing operations. Members of the same IT service provider are also likely to be spread across different sites and countries to be close to their customers. It is of utmost importance that communication channels be opened and maintained between those different 'local' teams so that the exchange of local experience is facilitated and best practice approaches to deliver similar world-class services throughout the aircraft manufacturer can be identified and implemented.

The following ways to *exceed customer expectations* have been suggested:

- demonstrate understanding of customers' expectations
- leverage the delivery dimensions
- exceed expectations of selected customers
- under-promise and over-deliver
- position unusual service as unique, not the standard.

Having looked at this comprehensive framework as to how superior service quality can be systematically created (the 'services marketing' perspective), we shall now turn to a concept of performance objectives, including their effects, as well as measurement and visualization of performance against these objectives over time ('management of operations' perspective).

The Five Performance Objectives

Let us now turn to the concept of performance objectives, as proposed by Slack, Chambers and Johnston (2004). This section covers the definition and presentation of each objective, as well as their internal and external effects, whereas the following section covers the measurement and visualization of performance over time along those performance objectives. We will do this from a service operation's viewpoint, using a range of service operations from the aviation context to illustrate and review the individual objectives.

Slack, Chambers and Johnston (2004) define a basic set of five performance objectives that apply to all operations – *quality*, *speed*, *dependability*, *flexibility* and *cost*. All these performance objectives have external and internal effects and are to a high degree interdependent. Using this set of objectives, an operation can clearly define the targets that it aims to achieve and also measure its performance along those targets over time. Obviously, it is most critical that those targets reflect what customers really want. Each performance objective will usually be split into several

Example: Luebeck Airport

A few years ago, Luebeck Airport was hardly heard of, being a small airport mainly used for private aviation. This has changed dramatically since Ryanair, the European low-cost pioneer, started offering cheap flights (first only) to London Stansted with their fleet of Boeing 737s. Very soon, major companies' employees started using Ryanair for their business trips between London and Hamburg.

Waiting area (Photo: Author)

Despite the limited size of the airport, all necessary facilities have been put in place in order to cope with the increasing numbers of customers and flights that are offered from Luebeck. For instance, the airport is using a massive marquee as a temporary terminal, comprising security, passport control, a bar, a duty-free shop, toilets and a children's play area.

Play area (Photo: Author)

Temporary terminal (Photo: Author)

In the case of a young family with small children that flies from Luebeck Airport, the customer flow process would typically run as follows. The family arrives at the airport by car, enters the short-term parking zone (restricted but short-term free parking), stops at the entrance door and unloads all their luggage. The family waits with the luggage in the airport building while one of the parents drives the car to the free long-term parking lot next to the airport (300–500m). Once reunited, they check in, look at the two shops in the main building and have something hot to eat in the very nice but cheap airport restaurant (with play area). After the meal, they stroll along to the security checkpoint and enter the temporary terminal (with bistro, duty-free shop and another play area) until the boarding.

As other low-cost airports, Luebeck Airport needs to stay attractive for low-cost providers by knowing and meeting their customer airlines' expectations and those of their passengers. Hence, service quality management over time must be one major concern to the airport.

relevant partial measures that are both easy to measure over time and meaningful to the operation in terms of what customers expect from it.

In Chapter 3, four different service operations from the aviation context will be considered to clarify what each individual performance objective may stand for in different situations. Those service operations are an airline (external service provider), an airport security company (external service provider), an IT support service within an aircraft manufacturer (internal service provider) and a spare-parts dispatch unit (both external and internal service provider). For those concrete examples, please refer to Tables 3.3 (quality), 3.4 (speed), 3.5 (dependability) and 3.6 (flexibility). Those examples should not be understood as recommendations that each service operation should implement, but they merely serve to explain what may be the perceived meaning of each performance objective to the customers of different service operations.

Quality

This can be defined as doing things 'right' or the extent to which products and services are delivered to specification. However, quality means different things in different operations (see Table 3.3 for concrete examples).

External (that is, towards external or internal customers), high quality satisfies customers. To put it the other way round, negative word-of-mouth communications amongst actual and potential customers is less likely to arise if quality is high, keeping in mind that negative word-of-mouth communications by irritated customers are most damaging to the success of an operation, as has been shown previously. But there are tremendous internal (that is, within the service operation itself) effects of high quality, too. The costs of actually producing the products and services provided are reduced if mistakes do not have to be corrected at a later stage. Furthermore, high quality increases dependability (that is, delivery of the service when and where promised), with positive effects on revenue, cash flow and again customer satisfaction. Finally, if quality is high internally, the level of mutual trust, internal dependability and stability of the operation will rise.

It can be summarized that high quality positively influences customer satisfaction and leads to stable and efficient operations.

Speed

This can be defined as short delivery lead times or how long customers have to wait for their products/services. But speed means different things in different operations, too (see Table 3.4 for concrete examples).

Externally, speedy delivery of services to customers (that is, short waiting times) will, in most cases, contribute to satisfy customers. Internally, speed reduces the need for inventory, shortens delivery lead times and therefore reduces risks, because the basis for decision-making is better if shorter-term forecasts can be used that tend to be more accurate.

It can be summarized that externally high speed contributes to satisfy customers and internally enhances high productivity via fast throughput.

Dependability

This can be defined as delivery at an agreed place and time, or when and where promised. However, also dependability means different things in different operations (see Table 3.5 for concrete examples).

Externally, dependability increases the level of trust customers have in the operation and their satisfaction with the services provided. Internally, operations that are dependable are more effective because of three contributions:

- First, dependability saves time, because not much time needs to be spent on sorting out alternative or replacement solutions.
- Second, money is saved because there are no extra efforts needed in terms of buying extra work, transportation or alternative services, to cover up for something not delivered where and when promised.
- Third, internal dependability leads to higher levels of mutual trust amongst the people contributing to produce the services provided. As a result of this, predictability and, therefore, stability of the internal processes increase and employees can focus on their immediate activities instead of having to worry permanently about what recovery actions need to be done.

In summary, dependability leads to customer satisfaction externally and high productivity internally.

Flexibility

This can be defined as being able to change the operation in some way to adjust it to change or take into account a new situation. Slack, Chambers and Johnston (2004) distinguish four different types of flexibility that may be relevant to an operation:

- *product/service flexibility* means to be able to introduce new product/services as the opportunity or need arises;
- *mix flexibility* means to offer a wide range of different products/services;
- *volume flexibility* means to be able to cope with different levels of throughput (qualities or volumes);
- *delivery flexibility* means the ability to change delivery times and/or places.

Also flexibility means different things in different operations (see Table 3.6 for concrete examples).

External, high flexibility is likely to increase the level of customer satisfaction. Flexibility inside the operation speeds up response, saves time spent waiting for an alternative solution and helps to maintain dependability; in other words, internally, flexibility contributes to high productivity because it represents the operation's ability to change.

Cost

This can be defined as the real value of investments necessary to provide a service. For some of those operations that compete directly on cost, the cost objective is the most important one. Still, no matter what the priorities of an operation are, the cost objective is almost always a universally attractive one.

Operations need to spend money or other resources on a specific mix of staff (such as the employees running an operation), facilities, technologies and equipment (such as PCs, licences, company cars, office rent, etc.), as well as bought-in material and services (such as postal services, call centre services, Internet services, etc.).

Externally, low costs allow for high margins or low prices or both. Internally, low costs allow for high productivity. Cost is affected internally by all other performance objectives (see Figure 2.13).

All performance objectives need some form of measurement as a prerequisite of systematic or controlled improvement. Slack, Chambers and Johnston (2004) define performance as the 'degree to which an operation fulfils the five performance objectives at any point in time, in order to satisfy its customers'.

One useful way to visualize both the importance of and/or the performance measured along the defined performance objectives is to use the polar representation of the performance objectives. This representation can also be used to contrast own against competitors' objectives or performance along those.

As an example, the relative importance of the five performance objectives of a spare-parts dispatch unit is contrasted with that of an airport security company in Figure 2.14. The spare-parts dispatch unit may focus primarily on dependability (delivery at an agreed place and time) and volume flexibility (being able to cope with low or high quantities of requests), followed by speed (fast delivery). Low prices to the customer and friendly and well-dressed staff may be considered to be of lower importance to this service operation.

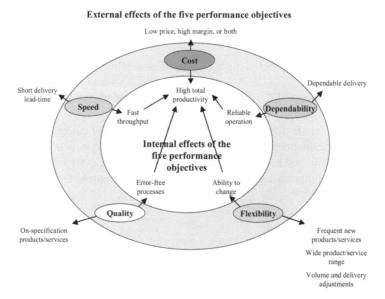

Figure 2.13 Internal and external effects of performance objectives
Adapted from Slack, Chambers and Johnston (2004)

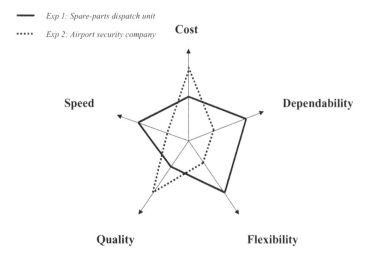

Figure 2.14 The polar representation of performance objectives
Adapted from Slack, Chambers and Johnston (2004)

The aircraft security company, on the other hand, may focus on low prices to the customer, since there is fierce competition in the security sector that is predominantly based on price, and good quality of its services in terms of effectiveness in finding weapons, explosives, etc. (This is critical since one single weapon that remains undetected is potentially an immediate threat to many people's lives.) Speed, dependability and (mix) flexibility may be considered to be of lesser importance.

Both the perceived importance of performance objectives and the performance measured by them are likely to change over time.

Having introduced the concept of performance objectives, the next section will present a very powerful method of how those performance objectives can be visualized against importance to customers and customer performance measurements against those objectives.

The Importance-Performance Matrix

This section presents the importance-performance matrix, as suggested by Slack, Chambers and Johnston (2004). This matrix is based on the concept of performance objectives described above, and a very powerful way to visualize what customers really want, to identify the relative importance of each need or want to them, and, at the same time, to find out how the operation is perceived by the customer to meet those expectations as compared to competitors.

Customers' wants and needs can be grouped into *order-winning, qualifying,* and *less important factors.* A nine-point scale each for the relative importances and the performance measured along those customer needs and wants has been proposed (see Tables 2.3 and 2.4).

Considering a public contracting agency as an example (such as for a non-privatized airport), price is likely to be seen as a qualifying factor because there is probably only a limited budget available. So if offers during the bidding process exceed (by far) the budgeted sum for a service to be contracted, those offers are likely to be rejected straight away. The same may be true for quality. If offers do not promise to deliver to specification they will be rejected. Speed and dependability, on the other hand, may be order-winners, if they are higher than competing offers. Flexibility, especially volume flexibility, will possibly be seen as a less important factor, as long as quality requirements are met.

Table 2.3 shows how the relative importance of needs and wants to the customer (or objectives from the operation's perspective) can be grouped and recorded.

Similarly, Table 2.4 shows how the perceived performance of the operation along those needs and wants or objectives can be grouped and recorded.

The importance-performance matrix helps to visualize the customer's perceived importance of individual performance objectives or service standards (that are defined from an operations perspective in order to meet customers' expectations), on one hand, and the operations performance against competitors as perceived by the customers, on the other hand. It can serve, therefore, as a valuable instrument in strategy development and a powerful basis for decision-making (see Figure 2.15).

The matrix shows four distinct fields or zones: the *appropriate zone,* the *improve zone,* the *urgent action zone* and the *excess zone.* If a measure lies in the appropriate zone, this indicates that it should be considered satisfactory. Performance must be higher for a standard or objective that is perceived as important by customers, but can be lower for those considered less important. If a measure lies in the improve zone, this indicates that improvement needs to be made. If a measure lies in the

Table 2.3   A nine-point scale of importance

Order winner	Strong	1	Provides a crucial advantage
	Medium	2	Provides an important advantage
	Weak	3	Provides a useful advantage
Qualifier	Strong	4	Needs to be up to good industry standard
	Medium	5	Needs to be up to median industry standard
	Weak	6	Needs to be within close range of the rest of the industry
Less important	Strong	7	Not usually of importance but could become more so
	Medium	8	Very rarely considered by customers
	Weak	9	Never considered by customers

Adapted from Slack, Chambers and Johnston (2004)

Table 2.4   A nine-point scale of performance

Better than competitors	Strong	1	Considerably better than competitors
	Medium	2	Clearly better than competitors
	Weak	3	Marginally better than competitors
Same as competitors	Strong	4	Sometimes marginally better than competitors
	Medium	5	About the same as most competitors
	Weak	6	Slightly lower than the average of most competitors
Worse than competitors	Strong	7	Usually marginally worse than most competitors
	Medium	8	Usually worse than competitors
	Weak	9	Consistently worse than competitors

Adapted from Slack, Chambers and Johnston (2004)

urgent action zone, this indicates that immediate steps should be taken to return to the appropriate zone. If a measure lies in the excess zone, this indicates that the operation is performing very well but in an area not valued or perceived as important by customers. The question has to be asked whether those resources devoted to achieving this high level of performance should not be deployed differently to achieve improvements where they are valued more by the customers.

In the following example of an applied importance-performance matrix, the performance of a low-cost airline is visualized in terms of customer perceptions of

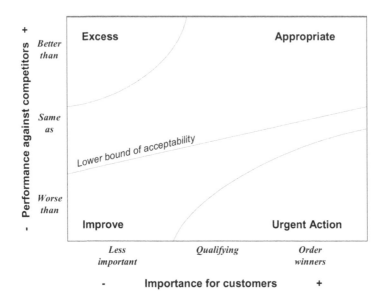

Figure 2.15 The importance-performance matrix
Adapted from Slack, Chambers and Johnston (2004)

key indicators compared to alternative, traditional airlines (see Figure 2.16). The example shown represents four concrete service standards that a low-cost airline service provider may consider as particularly relevant to control its service quality over time. In terms of the five performance objectives those standards are:

- partial measures of quality (inflight service quality and ordering process quality)
- dependability (punctuality of flights)
- cost (ticket price).

Other performance objectives, such as speed, or a corresponding service standard, like a turnaround time between flights of no more than 25 minutes, are certainly important as an internal performance objective but are not considered here, because it can be argued that passengers may not necessarily be directly involved or particularly interested.

The provider may define corresponding service standards as follows:

- *Ticket price* – Ticket prices on each route are always at least 50 per cent cheaper than comparable (average) ticket prices from traditional airlines and at least 10 per cent cheaper than those of other low-cost providers on the same routes.
- *Ordering process* – The ordering process on the Internet is very customer-friendly and straightforward, and most customers booking perceive the web-based ordering process as easier than that of competing airlines.

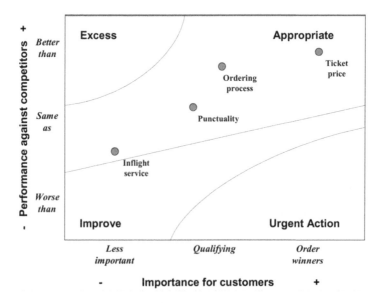

Figure 2.16 Example of the applied importance-performance matrix

- *Punctuality* – Most customers perceive the flights as just as punctual than those of traditional airlines.
- *Inflight service* – Most customers perceive the onboard service provided as satisfactory and flight attendants as friendly, helpful and competent.

The relative importance of the service standards defined above may have been established by asking customers using questionnaires, interviews and focus groups. The latter is a very powerful means of finding out what customers really want but there lies the danger of strong mutual influences between individual group members. Focus groups are rather better suited to collect qualitative inputs (hints, ideas, etc.) rather than exact quantitative measures.

The result of such activities may be that customers perceive the ticket price a clear order-winner, the ordering process and punctuality as qualifying factors, and inflight service as a less important factor.

The perceived performance of the low-cost airline against traditional competitors along those service standards should be measured regularly by using the same means as before. In the given case, the result may be that customers perceive the ticket price as clearly cheaper than competitors', the ordering process marginally easier and more customer-friendly than that of competitors, punctuality sometimes marginally better than that of competitors, and inflight service slightly worse than competitors'.

The position of the four points in the importance-performance matrix indicates that all lie in the 'appropriate zone' and no urgent actions are needed. The relatively poor rating of inflight service should be rather worrying to any traditional airline that predominantly serves long-distance routes. However, in the given case, customers have expressed that what matters to them is the low ticket price and inflight service is considered as less important. In other words, the low-cost airline gets away with poorer onboard service as long as the tickets are cheap.

Having introduced the importance-performance matrix as a means of visualizing the relative importance of performance objectives and an operation's performance against them, the next section will deal with change management issues.

Change Management Issues

As any change associated with the implementation of new methods or tools needs to be managed in some form in order to minimize harmful frictions, conflicts and resistance, some essential change management issues are discussed in this section. In doing so, we should first take a brief look at some developments in the general business context, before suggesting a number of strategies of how resistance can be pre-empted and how change can be managed and implemented within organizations.

Daniels (1998) argues that one fundamental idea in the 'formation for the corporation of the future is *systems thinking*. The idea is to view any business activity as a whole system of information, perception, values and activities.' She

Example: Ryanair

The European low-cost pioneer Ryanair started operations from and within Ireland in 1984 and has seen a number of dramatic changes since then. Currently, the airline is very successfully expanding in Europe, aiming to offer the lowest prices and the widest network of destinations. Ryanair is currently upgrading its fleet to the latest versions of the Boeing 737.

Ryanair boarding (Photo: Author)

The new seats used on board Ryanair aircraft are not foldable and have no pockets, saving money, weight, space and cleaning time. Safety instructions are glued on the back of the seats. If a flight is not fully booked, the free seats are kept empty in the front part of the aircraft to save fuel.

Ryanair interior (Photo: Author)

Advertisement space inside the aircraft is offered and used as an additional source of income. Food and even inflight entertainment in the form of portable DVD players are offered for cash during flights.

Temporary terminal used by Ryanair (Photo: Author)

Ticket-free booking via the Internet (and full payment in advance including fares, taxes and fees) certainly contribute to improve the cash flow situation of the airline. SKYTRAX only rated Ryanair a two-star airline in their Airline of the Year 2004 ranking. Financially, however, this does not mean that Ryanair is in trouble: the airline finished the business year 2004–2005 with amazing net profits of €268.9m.

Ryanair must be fully aware of what customers really expect and meet their requirements better than competitors do. The airline's mission, as has been repeatedly declared by its CEO Michael O'Leary, is to offer its customers Europe's lowest airfares, the best punctuality and customer service. Hence, service quality management is one major concern to the low-cost carrier.

further argues that 'as we move on from the industrial age to the information age, we are moving from a functional orientation of business to a process orientation'. In a process organization 'the entire company focuses on achieving processes that run across the organization, such as achieving high customer satisfaction, becoming more global, or restructuring the cost-base of the company'.

An example for such tendencies is Airbus, where the traditional division of the company in departments has been replaced recently by an organization of different Centres of Excellence that assume full responsibility for a specific, integrated part of an aircraft (for example, wings). Those Centres of Excellence are supported by the traditional functions (human resources, finance, procurement, etc.) as well as Centres of Competence (teams of specific experts), and they follow newly defined core processes across and beyond the company.

Also, Hammer and Stanton (2000) argue that some companies have successfully re-engineered their core processes by unifying activities that were previously divided into sub-activities done by different departments, and cutting off activities that do not add value to the overall process. Those companies assign outstanding managers as 'process owners', responsible for optimizing core processes across all traditional functions. They are given the authority to form multi-functional teams and assigned specific resources to achieve their objectives. In addition to that, new evaluation and reward systems are introduced, as well as new career programmes. What really matters to those 'process organizations' is real teamwork and the customer.

Change in general is not something that a company should try to suppress. On the contrary, change that is carefully managed can be argued to be necessary and beneficial to enable the company to cope with external and internal developments that cannot directly be influenced by the company. Companies need to constantly rethink whether their strategic goals are still in the best interest of the company and if what they are doing is still suitable to reach those strategic goals. Events such as the 9/11 attacks and market developments like the appearance of low-cost airlines, for instance, mark points or periods in time when change is most urgently needed. Many airlines, airports and also aircraft manufacturers – usually those that were best able to change – managed to survive and adapt to the new situation while others went out of business.

But also less dramatic changes, such as the implementation of a new service quality management method as suggested in Chapter 3, are potentially prone to obstacles and resistance. Change management issues need to be addressed to ensure that the advantages of the method are fully realized.

Major sources of implementation problems are confusion, role conflict and ambiguity, as well as conflict generation among employees, if the implementation is not carefully planned and carried out. Because considerable shifts in power and controllable resources occur in the course of most change implementation processes, some individuals and groups are likely to be jealous and show resistance to the structural changes. From a corporate culture perspective, change management becomes even more important. Hammer and Stanton argue that 'traditional leadership styles are inappropriate in process organisations. Managers cannot give orders and control; they have to negotiate and cooperate.' Processes should be standardized to a certain extent to improve productivity but leave the company sufficiently flexible to be able to customize offerings and satisfy differing customer needs.

A classic approach to dealing with resistance when implementing change has been offered by Kotter, Schlesinger and Sathe (1986). They identified four reasons for individual resistance to change in organizations:

- parochial self-interest
- misunderstanding and lack of trust
- different assessments of the benefits of change
- low tolerance for change.

They propose six strategies to deal with resistance depending on the given circumstances (see Table 2.5). Choices of those strategies have to be 'internally consistent and fit key situational variables.' The last two strategies of 'manipulation/ cooptation' or 'explicit/implicit coercion' should *not* be used, because long-term employee commitment to the changes are absolutely critical for any company to become a working process organization.

Brancheau and Wetherbe (in Gray et al. 1994) propose a number of steps to manage the introduction of new technology that are equally valid and can be extended to non-technology related changes as well:

- change agents should be used as early as possible;
- opinion leaders should be identified and targeted as early as possible;
- communication channels, especially informal ones, should be used;
- modelling and demonstrations should be used to convince opinion leaders;
- trial use helps reducing employee uncertainty and identifying areas for improvement at an early stage of implementation;
- local experts should be developed to provide support for long-term utilization.

Several alternative *implementation strategies* are widely used to implement new systems, but they can also be used to implement new methods and tools:

Table 2.5   Strategies for dealing with resistance

Approach	Commonly used where ...	Advantages	Drawbacks
1 Education/ Communications	... lack of information ... lack of analysis	+ once persuaded, people will help	– time-consuming
2 Participation/ Involvement	... initiator needs information ... others have power to resist	+ commitment + information and support	– time-consuming
3 Facilitation/ Support	... resistance due to adjustment problems	+ works best against problems	– time-consuming – expensive – can still fail
4 Negotiation/ Agreement	... initiator would lose ... group powerful resistor	+ easy way to avoid major resistance	– expensive – subordinate compliance
5 Manipulation/ Cooptation	... other tactics do not work ... or are too expensive	+ quick + inexpensive	– problems if people feel manipulated
6 Explicit/Implicit Coercion	... speed is essential ... initiator very powerful	+ speedy + overcomes any resistance	– risky if people get mad

Adapted from Kotter, Schlesinger and Sathe (1986)

- The *big bang* strategy, where the ready-made new system is implemented across the whole organization at once. Although duplication of work is saved using this method, it is very risky and represents a very high and sudden workload to employees.
- *Parallel running* is a lower-risk approach that allows for employees building up confidence with the new system because the old and the new way of doing things are running at the same time. However, this approach is expensive and means a higher workload for employees because there is duplication of work.
- The *phased approach* represents a useful compromise insofar as it allows for modifications and does not expose the whole company at once if there are still major problems with the new system. Within parts of the company the new system can be integrated by using either of the above approaches, until – step by step – the new system is implemented in the whole company.
- The *pilot study* approach seems suitable for high-risk projects, but is time-consuming and expensive.

Several studies have revealed that leaders play a crucial role in periods of change, be it for the implementation of new IT systems, organizational structures, working hours systems, or the like. Leaders have to foster change.

Goleman (1999) asked the question: 'What are the qualities that make someone an effective change catalyst?' He came to the conclusion that effective change leaders, beyond high levels of self-confidence and technical expertise, need high levels of 'influence, commitment, motivation, initiative, and optimism, as well as an instinct for organizational politics.' They have to initiate or manage change. People with this competence:

- recognize the need for change and remove barriers
- challenge the status quo to acknowledge the need for change
- champion the change and enlist others in its pursuit
- model the change expected of others.

Storey (1995) proposes an HRM (human resource management) model based on the beliefs that it is the human resource which gives a company the competitive edge, and that the aim should be real employee commitment to the company. He underlines, among other factors, the importance of top management involvement, employee empowerment, and culture management in the process of strategic change.

One very comprehensive and compelling approach to change management has been proposed by John P. Kotter (1996) with his 'eight-stage process of creating major change' in his book *Leading Change* (see Table 2.6). He argues that most changes suffer from sufficient momentum over time and tend to sink in the 'quicksand of complacency' or provoke serious resistance, because not all eight stages have been covered or only superficially so.

Table 2.6   The eight-stage process of creating major change

(1) Establishing a sense of urgency
- Examining the market and competitive realities
- Identifying and discussing crises, potential crises and major opportunities
- Fighting complacency (making crises visible; setting higher standards; re-structuring parts of the company in a way that widens employees' functional goals and allows for receiving customer contact; avoiding too much 'happy talk' by top management)

(2) Creating the guiding coalition
- Putting together a group with enough power to lead the change (position power, expertise, credibility and leadership as well as management skills)
- Getting the group to work as a team (creating trust and common goals)

(3) Developing a vision and strategy
- Creating an effective vision to help direct the change effort (this vision needs to be imaginable, desirable, feasible, focused, flexible and communicable)
- Developing strategies for achieving that vision

(4) Communicating the change vision
- Using every vehicle possible to constantly communicate the new vision and strategies (keep it simple; use verbal pictures and all channels available; especially emphasise dialogue whenever possible; keep repeating the message; lead by example; explain seeming inconsistencies)
- Having the guiding coalition role-model the behaviour expected of employees

(5) Empowering employees for broad-based action
- Getting rid of obstacles (that is, barriers to empowerment such as bosses discouraging employees from participating in the change process, formal structures making it more difficult for employees, lack of needed skills, lack of suitable information systems)
- Changing systems or structures that undermine the change vision
- Encouraging risk-taking and non-traditional ideas, activities and actions

(6) Generating short-term wins
- Planning for visible performance improvements (wins)
- Creating those wins
- Visibly recognizing and rewarding the people who made those wins possible

(7) Consolidating change and producing more change
- Using increased credibility to change all systems, structures and policies that do not fit together and do not fit the transformation vision
- Hiring, developing and promoting people who can implement the change vision
- Reinvigorating the process with new projects, themes and change agents

(8) Anchoring new approaches in the culture
- Creating better performance through customer- and productivity-oriented behaviour, more and better leadership, and more effective management
- Articulating the connections between new behaviours and organizational success
- Developing means to ensure leadership development and succession

Adapted from Kotter (1996)

Kotter emphasizes that there is a crucial and clear distinction between management and leadership that must be understood in order to avoid misunderstandings about managers' roles in a company and get the 'right' people to do the matching job in major change processes.

'Management', according to Kotter, means planning, budgeting, organizing, staffing, controlling and problem-solving, and produces a degree of predictability and order. It has the potential to consistently produce the short-term results expected, such as being on time and within budget.

'Leadership', on the other hand, means establishing direction (developing a vision of the future), aligning people, motivating and inspiring. Leadership has the potential to produce extremely useful change (often to a dramatic degree), such as new products that customers want, new organizational structures that are more responsive and focus on core processes as opposed to pure functionality, or new approaches with labour unions that make a firm more competitive in the marketplace.

Leadership creates a vision and strategies to achieve that vision, whereas management creates plans to implement those strategies and budgets to convert those plans into financial projections and goals.

This is not to say that a manager should either manage or lead but rather he/she may have to do both, depending on his/her position and responsibilities. In general, top management should focus more (but not exclusively) on leadership (to give direction), whereas middle and lower management should focus more (but not exclusively) on management (to make it happen). The crucial point here is to use the right mix of resources at each level to provide both good leadership and at the same time sound management within any company.

To summarize, it can be said that carefully planned and managed change is more likely to result in the successful implementation of a new way of working. The above offers some advice on what needs to be considered and which approach could be taken in a specific situation. The method proposed in Chapter 3 also covers change management issues based on the above theories.

Quality in Aviation

There are a high number of formal quality management approaches across different industries that are also applied within the aviation context. This section covers some general aspects of quality management and takes a closer look at some of the most important examples of quality standards, prizes and awards, as well as rankings and reports.

Quality, as was stated above, can be defined as 'doing things right' in terms of meeting customer expectations from a product or service itself – irrespective of its cost, speed of delivery, reliability of the provider (the reliability of the product or service itself would be one aspect of its quality) and flexibility of the provider.

Since customers are getting increasingly demanding about the quality of the products and services they buy, achieving quality that is at least up to an industry's standard should be a major concern for any company. This is even more so when safety and security issues are involved and customers have to pay a lot of money to buy the products and services in question, as is the case in the aviation industry. Flight tickets are often quite expensive and customers expect corresponding service quality for their money, although it can be argued that most low-cost passengers would not expect the same service levels. Very high safety and security-related quality levels are obviously a must in either case and customers would not be willing to compromise on this aspect.

Since achieving and maintaining high quality levels in the aviation industry is very difficult and challenging, due to its complexity and tough safety and security requirements, the implementation and use of a quality management system is essential; for instance, to define quality objectives, measure results and allow for continuous improvement of quality over time.

General Aspects

For companies above a certain size, formal and standardized quality management systems, such as defined by the ISO 9000 series, not only have tremendous potential to achieve the aim of improving quality over time, but the certifications or awards received in the process of implementation clearly serve as a marketing asset and can be shown to existing and potential customers in order to convince them of the quality of the company's products and services.

Specific benefits of introducing or maintaining such a quality management system are that:

- the company's policies and objectives are set and signed off by 'top management';
- understanding of customers' requirements with a view to achieving customer satisfaction is enhanced;
- internal and external communications are improved and formalized;
- understanding of the organization's processes is enhanced;
- understanding how statutory and regulatory requirements impact on the organization and their customers is enhanced;
- clear responsibilities and authorities are agreed for all staff;
- the use of time and resources is improved;
- wastage is reduced;
- consistency and traceability of products and services are enhanced;
- morale and motivation is improved.

The service quality cycle proposed in this book (see Chapter 3) helps to obtain certification to standards, increases the likelihood of being awarded quality-related prizes and awards, and helps to achieve better ranking results in relevant surveys or better standings in specific reports.

Standards – The ISO 9000 Series

The ISO 9000 series are the most commonly used international standards that provide a framework for an effective quality management system. The ISO 9000 family of standards was revised in 2000 (ISO 9000:2000 series). So far, they have been implemented in some 634,000 organizations in 152 countries and can be said to have become an international reference for quality management requirements.

To be fair, it must be said that actual certification according to the ISO 9000 series may not be worthwhile in the case of smaller companies unless there is a clear requirement to be certified, because the efforts needed and costs linked to becoming certified are relatively high. However, the principles of the ISO 9000 series represent a valuable framework.

In order to gain maximum benefits from the ISO 9000:2000 series, a number of steps should be taken:

1. It should be defined why the organization is in business.
2. The key processes that state what is done by the company should be determined.
3. It should be established how those processes work within the business.
4. The owners of these processes should be determined.
5. These processes should be agreed throughout the organization.

There are a number of different standards in the ISO 9000:2000 family, the most relevant ones in the context of this book can be argued to be the following:

- ISO 9000 Quality Management Systems fundamentals and vocabulary installation and servicing
- ISO 9001 Quality Management Systems requirements (the requirement standard)
- ISO 9004 Quality Management Systems guidelines for performance improvement.

ISO 9001:2000, the requirement standard, covers the following sections:

- Quality Management System
- Management Responsibility
- Resource Management
- Product Realization
- Measurement Analysis and Improvement.

The standard ISO 9001 not only covers the requirements in order to define and install a quality management system, but also requires that you plan and manage the processes necessary to continuously improve it over time.

The service quality cycle as proposed in Chapter 3 can help in two ways: Either, it can be used by a company as a method within the ISO 9000 series quality management system to achieve certification or maintain it. Alternatively, for companies that do not wish to become certified but still want to follow the principles of the ISO 9000 series, it can be used as the method to apply in order to manage and continuously improve service quality over time.

Prizes and Awards

There are many international and national quality awards and prizes, most of them suggesting a clear framework of what needs to be done in order to qualify for such an award or prize. Just a few examples are listed here:

- US State Quality Awards, such as the Arizona State Quality Award
- The Malcolm Baldrige National Quality Award (US)
- The Deming Award for Quality (Japan)
- The European Quality Award
- The President's Quality Award (US).

As an example of one tremendously successful quality award programme, the Baldrige National Quality Award of the US will be discussed in more detail.

The Baldrige National Quality Award   On 20 August 1987, President Ronald Reagan signed the 'Malcolm Baldrige National Quality Improvement Act of 1987'. This act can be argued to have helped in making quality a national priority and revitalizing the US economy in the 1990s. More than 43 US states and many countries, including Japan, have quality programmes modelled after Baldrige. In particular, the Baldrige 'Criteria for Performance Excellence' (see Table 2.7) are widely used as a framework for assessment and improvement.

Malcolm Baldrige served as US Secretary of Commerce from 1981–1987, when he tragically died in a rodeo accident. His reputation is that of an excellent manager who contributed to long-term improvements in efficiency and effectiveness of government.

The US Department of Commerce is responsible for the Baldrige National Quality Program and the Award. The National Institute of Standards and Technology (NIST), an agency of the Department's Technology Administration, manages the Baldrige Program.

Many companies do not necessarily apply for the Award with the idea of winning it, but with the goal of receiving the evaluation of the Baldrige examiners. The main benefits for companies that participate in the programme can be argued to be in the process itself, whether or not they receive the Award. Those, however, who do win it, will usually receive it from the President of the US and the Award can certainly be considered a major marketing asset and advantage for the company.

Example: Munich Airport

Since May 1992, when Munich Airport Franz Josef Strauß was inaugurated and took over from the more central but very confined Munich Airport Riem, the new airport has developed remarkably well and is now, in terms of passengers, number eight in Europe and number two in Germany after Frankfurt.

Munich Airport from above (Photo: Munich Airport)

In terms of domestic passengers, with 8.78 million in 2004, it is even number one in Germany in front of Frankfurt. The airport's shareholders are the State of Bavaria (51 per cent), the Federal Republic of Germany (26 per cent) and the City of Munich (23 per cent).

Munich Airport is located 28.5 km northeast of the centre of Munich but is fully integrated into the regional public transport system. There is an underground S-Bahn rail station near the terminals. Also, a motorway was extended especially to ensure easy and quick access to the airport. The fact that there are 20,000 parking spaces available for customers makes access by car even more convenient.

ATR 42 and Tower (Photo: Munich Airport)

Passenger area (Photo: Munich Airport)

With its two big terminals, the airport has a total capacity of over 45 million passengers per year. There are 149 aircraft parking positions, 61 boarding stations, two parallel runways, and a baggage conveyor system of a total length of 58 km that is able to handle 33,200 pieces of luggage per hour. Currently, only just under two thirds of this capacity is used.

Munich Airport and a Lufthansa B737 (Photo: Munich Airport)

Munich Airport successfully completed the ISO 9001:2000 certification process in January 2002. In addition to that, following the SKYTRAX Best Airport 2004 Survey, Munich was rated airport number three in Europe in terms of overall service and product quality.

The service quality cycle that is proposed in Chapter 3 of this book clearly has the potential to directly contribute to improve a company's standing against criteria 2.1, 3.1, 3.2, 4.1, 5.2, 7.1 and 7.2 of this framework (see Table 2.7).

Specific Rankings and Reports

In the aviation context, there are many national and international rankings and reports that cover almost all aspects of airline and airport performance, but increasingly aircraft manufacturers' performance too, although the latter is mainly focused on reliability and accident rate perspectives. In the following, two important examples will be considered – SKYTRAX international airline and airport rankings, and the Air Travel Consumer Report that covers different aspects related to US airlines and airports.

SKYTRAX   Since 2000, SKYTRAX a London (UK) based independent research organization, operates the Star Rating System for the world airline industry and

Table 2.7   Criteria for performance excellence

Categories and items	Point values
1 Leadership	**120**
1.1 Organizational leadership	70
1.2 Social responsibility	50
2 Strategic planning	**85**
2.1 Strategy development	40
2.2 Strategy deployment	45
3 Customer and market focus	**85**
3.1 Customer and market knowledge	40
3.2 Customer relationships and satisfaction	45
4 Measurement, analysis and knowledge management	**90**
4.1 Measurement and analysis of organizational performance	45
4.2 Information and knowledge management	45
5 Human resource focus	**85**
5.1 Work systems	35
5.2 Employee learning and motivation	25
5.3 Employee well-being and satisfaction	25
6 Process management	**85**
6.1 Value creation process	50
6.2 Support processes	35
7 Business results	**450**
7.1 Customer-focused results	75
7.2 Product and service results	75
7.3 Financial and market results	75
7.4 Human resource results	75
7.5 Organizational effectiveness results	75
7.6 Governance and social responsibility results	75
Total points	**1000**

Adopted from www.baldrige.com

similarly now also for the international airport industry, based on international service standards (where applicable).

SKYTRAX airline rankings   The SKYTRAX Airline of the Year Survey is the world's largest passenger survey; for instance, making over 10.8 million entries in the 10-month period from June 2003 till March 2004. The survey is conducted by SKYTRAX Research of London (UK) and there is no external sponsorship or influence as to any aspects of the survey.

In the 2004 Airline of the Year Survey, passengers from over 92 countries participated in:

- online customer surveys
- SKYTRAX Travel Panel
- corporate travel questionnaires and interviews
- consumer telephone interviews
- selective passenger interviews.

Additional research was conducted using detailed back-up interviews with representative samples of survey respondents. No financial payment was made to any of the respondents, the cost being entirely funded by SKYTRAX. The only material incentive was that all respondents participated in an electronic draw and were given the chance to win two long-haul tickets with the previous year's Airline of the Year (the tickets were supplied by SKYTRAX).

In the following, the meaning of an airline being awarded five, four, three, two or one stars is presented:

***** Five-Star Airlines
The ultimate approval that is only awarded to airlines supplying the highest quality performance. This ranking recognizes airlines at the forefront of product innovation that generally set trends and standards that will be followed by other carriers.

**** Four-Star Airlines
Approval that is awarded to airlines supplying a good quality performance. Four-star airlines provide a good standard of product across all travel categories and of staff service delivery onboard and in the airport environment.

*** Three-Star Airlines
This grading is awarded to airlines supplying a fair quality performance of about industry average. This ranking signifies a satisfactory standard of product across most travel categories but may reflect less consistent standards of staff service either onboard and/or in the airport environment.

** Two-Star Airlines

Grading for poor airline quality performance that falls below the industry average in each competitive market sector. Two stars represent a poor standard of product across different ranking categories or poor and inconsistent standards of staff service delivery onboard or in the airport environment.

* One-Star Airline

Grading for airlines showing very poor quality performance with very poor standards of product across all travel categories and poor or inconsistent standards of staff service delivery both onboard and in the airport environment.

Table 2.8 shows the main criteria headings that are used as a framework to rate each airline. The results in all of the categories of criteria are then translated into one overall star rating for each company. Criteria that are not applicable for individual airlines are not considered in those specific cases. For instance, if one airline only serves short-haul routes with a single class layout, this airline will not be rated against first class, business class and long-haul criteria of the framework.

The examples in Table 2.9 show selected survey results from 2004 (the latest rankings for 2005 have been announced in the course of publication of this book). In 2004, Singapore Airlines was rated the best airline of the world. In contrast, three European low-cost carriers are considered, Ryanair, easyJet and dba.

SKYTRAX airport rankings Similar to the airline rankings described above, SKYTRAX also conducts surveys and rankings on airports worldwide. In doing so, particular emphasis is put on those airports' service standards as a whole, also covering immigration, restaurants, retail outlets, etc. – even though it is recognized that each sector of product or service may come under a different quality ownership and control. This is done because customers tend to integrate all their perceptions into one impression that is linked to the airport as a whole.

In the course of the survey, passengers from all over the world are asked to assess airport standards on the key features of:

- ground transportation, location, signposting, access
- ease of transfer
- passenger facilities
- staff service, staff availability
- terminal comfort
- security/immigration.

From June 2003 till March 2004, for instance, passengers from 86 countries made over 4.8 million entries. The airport rankings are sub-divided into regional rankings.

Table 2.8   Criteria for airline ranking

Summary ranking
Combined quality of product/staff service delivered to passengers in airport and onboard environments
    First Class
    Business Class
    Economy Class
Airport services
    Check-in – First/Business
    Check-in – Economy Class
    Transfer services – First/Business
    Transfer services – Economy Class
    Arrival services
    Business Class lounge – product facilities
    Business Class lounge – staff service
    First Class lounge – product facilities
    First Class lounge – staff service
Onboard features
Inflight entertainment may vary according to aircraft type
    Cabin safety procedures
    Inflight entertainment
    Passenger comfort amenities
    Onboard reading materials
Cabin seat comfort
Seating may vary according to aircraft type
    First Class – long-haul
    Business Class – long-haul
    Business Class – short-haul
    Economy Class – long-haul
    Economy Class – short-haul
Onboard catering
    First Class meals – long-haul
    Business Class meals – long-haul
    Business Class meals – short-haul
    Economy Class meals – long-haul
    Economy Class meals – short-haul
Cabin staff service
    First Class – service efficiency
    First Class – attitude/friendliness
    Business Class – service efficiency
    Business Class – attitude/friendliness
    Economy Class – service efficiency
    Economy Class – attitude/friendliness
    Responding to passenger requests
    Cabin presence through flights
    Assisting parents with children
    Staff language skills

Adapted from www.skytraxsurveys.com

Table 2.9   Examples of airline rankings

	Singapore Airlines	Ryanair	easyJet	dba
Summary ranking – *Combined quality of product/staff service delivered to passengers in airport and onboard*				
First Class	*****	-	-	-
Business Class	*****	-	-	-
Economy Class	****	**	***	***
Airport services				
Check-in – First/Business	*****	-	-	-
Check-in – Economy Class	****	**	***	***
Transfer services – First/Business	****	-	-	-
Transfer services – Economy Class	****	-	-	-
Arrival services	***	*	**	**
Business Class lounge – product facilities	****	-	-	-
Business Class lounge – staff service	****	-	-	-
First Class lounge – product facilities	****	-	-	-
First Class lounge – staff service	*****	-	-	-
Onboard features – *Inflight entertainment may vary according to aircraft type*				
Cabin safety procedures	****	***	***	***
Inflight entertainment	*****	-	-	-
Passenger comfort amenities	****	*	**	**
Onboard reading materials	****	-	-	**
Cabin seat comfort – *Seating may vary according to aircraft type*				
First Class – long-haul	*****	-	-	-
Business Class – long-haul	*****	-	-	-
Business Class – short-haul	****	-	-	-
Economy Class – long-haul	***	-	-	-
Economy Class – short-haul	***	**	***	***
Onboard catering				
First Class meals – long-haul	*****	-	-	-
Business Class meals – long-haul	****	-	-	-
Business Class meals – short-haul	****	-	-	-
Economy Class meals – long-haul	****	-	-	-
Economy Class meals – short-haul	****	* (pay bar)	** (pb)	*** (pb)
Cabin staff service				
First Class – service efficiency	*****	-	-	-
First Class – attitude/friendliness	****	-	-	-
Business Class – service efficiency	*****	-	-	-
Business Class – attitude/friendliness	****	-	-	-
Economy Class – service efficiency	****	**	***	***
Economy Class – attitude/friendliness	****	**	****	***
Responding to passenger requests	****	**	***	***
Cabin presence through flights	*****	-	-	-
Assisting parents with children	****	**	**	**
Staff language skills	****	**	***	***

Adapted from www.skytraxsurveys.com

Example: Singapore Airlines

Singapore Airlines is one of the leading service quality providers in the airline industry worldwide. As of June 2005, its fleet of 89 aircraft (27 B747-400s, 57 B777s and 5 A340-500s) is one of the youngest in terms of aircraft age, 5 years and 4 months on average. On 28 June 2005, Singapore Airlines established another world record by inaugurating the world's longest non-stop commercial flight with a daily service between Singapore and New York, using the new A340-500 aircraft. Doing so, the airline directly linked South-East Asia with New York for the first time ever. The airline's yearly growth in terms of passenger kilometres is currently 5.8 per cent.

From 2000–2004, Singapore Airlines has won a total of at least 262 international awards and prizes in categories such as Quality of Services, Best Crew on Board, Best Staff on Ground, Best First Class Service, Best Economy Class, Best Business Class, Best in Friendliness, Best in Cabin Outfit, Best in Service, Number One for Customer Service, Best Frequent Flyer Programme, Best Innovator, Best Leisure Airline, Best Service Onboard, Best Comfort, Best Airline in the World, Best Managed Company, Airline of the Year, Best Cabin Crew Service, Best Inflight Entertainment and Best for Punctuality and Safety, just to name a few. Also, Singapore Airlines was awarded the SKYTRAX Best Airline of the Year 2004 Award, followed by Emirates, Cathay Pacific, Quantas, Thai and British Airways.

Singapore Airlines has shown a number of highly innovative approaches to improving customer comfort and inflight entertainment. Some of the more recent examples are that passengers are offered interactive language courses during the flight (in cooperation with the Berlitz language school). Also, the airline offers the world's largest inflight choice of video games (85 in June 2005).

Singapore Airlines A340-500 (Photo: Airbus)

In the business year ending 2005, with revenues of over $12 billion (+14 per cent over 2003 and +23 per cent over 2004) as well as net profits of nearly $1.4 billion (+31 per cent over 2003 and +64 per cent over 2004), Singapore Airlines stands remarkably well financially and clearly demonstrates that world-class service quality goes with world-class profits.

Air Travel Consumer Report   This is a monthly product of the US Department of Transportation's Office of Aviation Enforcements and Proceedings (OAEP). Divided into five sections, the report covers:

- flight delays
- mishandled baggage
- over-sales
- consumer complaints
- customer service reports to the Transportation Security Administration.

Relevant statistics covering almost exclusively the US market are made available in the report. If, as an example, somebody wanted to compare the relative number of complaints made by passengers in a certain period of time about specific traditional airlines as opposed to a representative low-cost carrier, he or she could find the following: in March 2005, DELTA, United and American Airlines received 0.89, 0.9 and 0.93 complaints respectively per 100,000 enplanements as opposed to 0.16 received by Southwest Airlines.

The service quality cycle that is proposed in Chapter 3 as a method of implementing and running an effective yet efficient service quality management system is argued to have the potential to significantly improve service quality as perceived by customers. Hence, customers participating in interviews, questionnaires, etc. (that are leading to ranking positions or standings in specific reports) will give more positive feedback and therefore influence the results of these surveys correspondingly.

This chapter discussed a number of proven and essential theories and frameworks with specific emphasis on the work of Zeithaml and Bitner, Slack, Chambers and Johnston, as well as Kotter, that are internationally accepted as state-of-the-art in the fields of services marketing, management of operations and change management respectively. Also, quality-related aspects in aviation were considered, including some of the most important examples of standards, prizes and awards, as well as rankings and reports.

The next chapter describes a practical step-by-step method, based on these theories and frameworks, for how any service provider in the aviation context can implement an effective yet efficient service quality management system.

Chapter 3

The Service Quality Cycle

Chapter Summary

After the theoretical considerations presented in the previous chapter, this chapter describes a practical step-by-step method (the 'service quality cycle') by which any service provider can implement – with little effort – an effective system to manage service quality over time. This can be achieved by finding out and measuring what matters most to customers, formulating customer-driven service standards, measuring how the service operation is perceived by customers to perform along those valued standards over time, and what concrete actions can be derived from the results of such measurements in order to improve service quality. The chapter ends with a summary of benefits inherent in using this method.

Overview

The service quality cycle proposed in this chapter (see Figure 3.1) is meant to serve as a generic framework that is offered along with pieces of advice and hints as to how it should be used, but that needs to be adjusted to individual service operations, depending on their specific context. In other words, specific individual choices will have to be made by the users of the method. Depending on the choice of the users, the proposed method can be used with little effort to gain valuable yet rough indications; a lot of effort will gain more rigorous results and enable serious analyses to be conducted in very much more detail.

Method

This comprises six practical steps that need to be carried out regularly:

- finding out what customers expect from the service operation and translating those expectations into service standards (Step 1)
- finding out what is the relative importance of each identified service standard as perceived by the customer (2)
- measuring performance against those standards (3)
- visualizing and analysing the current situation (4)
- deriving action items based on this analysis (5)
- implementing corresponding changes and controlling the cycle over time (6).

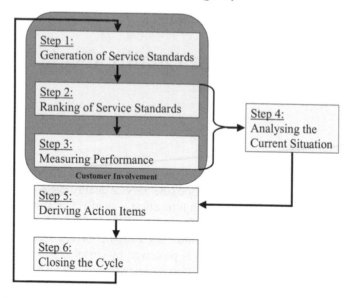

Figure 3.1   The service quality cycle

Depending on the specific circumstances of the organization concerned, some of the steps do not have to be carried out every time the cycle is gone through, or perhaps this may be done only at a lesser intensity, because the organization's own efforts and customer disturbances ought to be kept to a minimum in order to increase efficiency.

It is strongly recommended to have some form of customer involvement during Steps 1–3. In the following sections, each of the steps is explained in detail and suggestions offered as to how each step may have to be adjusted to a given operation. At the end of each section, more specific additional advice is offered for service providers within airline, airport and aircraft manufacturing operations that is related to one of the relevant key issues from the respective step of the method.

Step 1 – Generation of Service Standards

The first step (see Figure 3.2) is all about generating the right service standards that must reflect what customers really want from the service operation in question. If one does not know what customers want, the likelihood of not meeting their requirements is very high. Customer-driven service standards, therefore, represent the very basis of operational and – perhaps even more importantly – strategic decision-making. Therefore, it is essential to get them right.

In order to achieve the objective of formulating a set of customer-driven service standards that can and ought to be communicated internally to employees and externally to customers, a number of sub-steps are proposed here (see Figure 3.3).

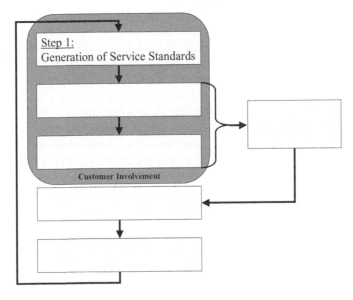

Figure 3.2 Generation of service standards

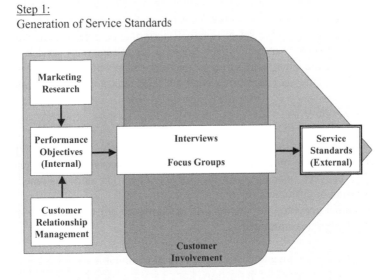

Figure 3.3   Step 1 – Generation of service standards

Step 1 (that is, the generation of suitable, meaningful and relatively easy-to-measure service standards that immediately reflect the major customer needs) should be taken once during the implementation of the service quality cycle. Those service standards should then be reviewed periodically, depending on the volatility of those customer needs and the overall situational context that may require strategic adaptation of the service operation over time (see Step 6).

It is critical to find the right balance of customer involvement, not annoying customers with constant requests (questionnaires, interviews, etc.) but still getting enough vital information to be able to put the service operation in line with customer needs. Many customers tend to see their involvement in improving the services provided as a pointless waste of their time. Yet, in order to better meet customer requirements, those services need to reflect real customer needs that can best be identified with the very help of customers. Therefore, customers have to be convinced that their time spent involved in the process is by no means wasted, but that, on the contrary, their efforts helping the researchers to generate the right standards allow the latter to put their organization's service efforts in line with customer needs. Hence, customers directly benefit from this joint effort.

The starting point of Step 1 is to look at essential marketing research and customer relationship management issues (that help to understand customer expectations). Based on those and the concept of the five performance objectives, a set of suitable (internal) performance objectives should be formulated that are validated, keeping in mind assumed customer needs. Once such a mature set of performance objectives has been defined, it should be presented to (key) customers for further validation ('customer involvement'). Based on the set of (internal) performance objectives as well as the inputs obtained during customer involvement, a set of (external) service standards should be generated that will serve as the service operation's new performance objectives (internally) and ought to be communicated to customers (externally).

Marketing Research

Understanding of customer expectations can be greatly enhanced by properly conducting and using marketing research. One key success factor is to emphasize interaction between managers of the service operation and customer contact employees and/or customers.

The internal generation of meaningful performance objectives can be enhanced tremendously by measures such as:

- continually tracking and keeping record of complaints
- applying best practice in the provision of services offered
- ascertaining present and future customer requirements
- finding out why customers ceased using the services offered.

How exactly this sort of marketing research should be conducted certainly depends – amongst other things – on the size and type of the service operation in question,

Table 3.1   Checklist – marketing research

- Actively find out the reasons for customer complaints.
- Record those reasons.
- Actively find out why customers have ceased using the services.
- Record those reasons.
- Actively find out the best practices in delivering the services offered.
- Record those best practices.
- Actively find out present and future customer requirements.
- Record those customer requirements.

the situational context (such as budget available, legal aspects, etc.), the number of customers that use the services offered, as well as their geographical locations. It could range from simply talking to customer contact employees about these issues and recording the information gained on a piece of paper to using complex marketing research methods and tools. In most cases, it would probably be sufficient to simply use a spreadsheet (such as Excel) to record the information gained by talking to key customer contact employees about the issues mentioned. If you are alone (that is, you are the only customer contact 'employee'), just record your own experience and the information that you gain from directly asking customers or receiving complaint letters, etc. If the number of customers and employees is higher, however, one would instead need some sort of customer database that everybody in the service operation permanently works with and makes entries immediately after or even during (telephone) customer contacts.

The important thing is to be able to have an overview of what caused customers to cease using the services offered, how specific tasks can be done best (best practice), what are causes for complaints, as well as some background information as to the specific context of each piece of information recorded. If it is recorded, for instance, that one customer left because there was nobody who could help with a specific problem, an additional piece of information might be that the expert employee was ill and there was no backup who could have replaced him or her.

Table 3.1 summarizes the recommendations relating to marketing research.

Customer Relationship Management

As identified in Chapter 2, the key factors related to building good customer relationships are:

- proper market segmentation (focusing on specific target segments)
- focusing on relationships rather than transactions
- focusing on existing customers as opposed to new customers.

Good customer relationships are worth investing in because it is cheaper to keep existing customers than to gain new ones. Also, satisfied long-term customers are likely to spread positive (highly influential) word-of-mouth reports about the services provided, both leading to increased profits. Furthermore, essential feedback from customers is likely to be more easily available if close customer relationships are maintained. For all these reasons, customer retention is of utmost importance. By monitoring relationships with key customers and regularly talking to both customers and the service operation's own customer contact personnel, beneficial long-term relationships can be positively influenced and negative developments anticipated.

By looking at recovery situations (when things go wrong) as opportunities to improve the service operation, and tracking and anticipating such recovery opportunities, the percentage of disappointed customers who cease using the service operation will decrease considerably. The impression must be created in the mind of the customer that, although something went wrong, the service provider made up for it, and quickly. Disappointed customers who can be turned around with appropriate recovery actions by the service operation will, in most cases, even become a source of positive word-of-mouth communications to other customers or potential customers. This is so because the service operation has dealt well with the situation, and customer expectations – as to how the situation was dealt with – may even have been exceeded.

Since speed is a critical matter in any recovery action, customer problems with the services provided have to be taken care of immediately, on the front line and by the contact personnel who first get to know of the problems from a disappointed customer. This is only possible if front-line (or customer contact employees) are actually empowered in a sufficient way that they are able to take the necessary steps, such as immediately after hearing about the problem in the presence of the angry customer. Obviously, the front-line employees will in most cases not be able to do all the work associated with a necessary recovery but what he or she does have to do is to immediately clarify with the person who can do this, so that the customer sees straightaway that his or her problem is taken care of seriously and without delay.

The minimum action to take is to trigger the recovery actions needed and making the customer aware of this fact. Also, an apology is useful, even if the problem was not caused by the front-line employee personally, because it underlines that the front-line employee feels a part of the overall service operation and regrets the fact that the customer has got a problem with the services delivered.

If the customer is given the impression that the front-line employee does not care personally or, worse, tries to blame the customer, he or she will turn away even more angrily and is likely to spread negative word-of-mouth reports that are very damaging to the service operation.

Learning from recovery experience and sharing this experience with all fellow front-line employees and managers of the service operation is crucial to improve the overall capability of the operation to deal with complaints.

Finally, customers must feel that they are truly appreciated and respected at *each* encounter, both as individual persons and as representatives of the customer organization. It is important that this mindset is 'lived' and meant seriously by all employees and management of the service operation, because this clearly and directly has an influence on customer relationships. This is *not* to say that the customer is

Table 3.2   **Checklist – customer relationship management**

• Monitor relationships with key customers.
• Keep talking with customers and own customer contact employees.
• Track and anticipate recovery opportunities.
• Take care of customer problems on the front line.
• Solve problems quickly.
• Empower front-line employees to solve problems.
• Record and learn from recovery experiences.
• Honestly appreciate and respect each individual customer and show it.

always right. Sometimes customers have to be told 'No', but they must be told in a respectful manner, stating the reasons why (perhaps because one particular service was not included in the contract or service agreement). Also, alternatives should be offered to the customer as to how the 'No' could be turned into a 'Yes'.

Table 3.2 summarizes the recommendations related to customer relationship management.

Performance Objectives (Internal)

With the marketing research and customer relationship management considerations in mind, a set of performance objectives need to be defined that are realistic, in line with the overall strategy of the company and, above all, reflect the needs and requirements of the customer. It is recommended here to use the framework of the five performance objectives – *quality, speed, dependability, flexibility* and *cost* – to formulate those objectives. All these basic performance objectives have external and internal effects and are to a high degree interdependent. Using this set of objectives, an operation can clearly define the targets that it aims to achieve and also measure its performance along those targets over time.

It is most critical that those targets reflect what customers really want. Each performance objective will usually be split into several relevant partial measures that are both easy to measure over time and meaningful to the operation in terms of what customers expect from it.

In the following tables, four different service operations from the aviation context will be considered to clarify what the quality, speed, dependability and flexibility performance objectives may stand for in different situations. Those example service operations are an airline (external service provider), an airport security company (external service provider), an IT support service within an aircraft manufacturer (internal service provider) and a spare-parts dispatch unit (both external and internal service provider). The examples do not represent recommendations that each service operation should implement; they merely serve to explain what may be the perceived meaning of each performance objective to the customers of different service operations.

Table 3.3   'Quality' means different things in different operations

Example service operation	What does 'quality' mean from a customer's viewpoint?
Airline (external service provider)	Clean aircraft Comfortable seats Good selection of newspapers and magazines Friendly and competent cabin staff Standard foods and drinks offered on board Safety equipment appears to be in good state Adequate air temperature Good inflight entertainment Clean toilets No bumpy landings
Airport security company (external service provider)	No long queues to go through security Friendly (yet determined) security staff Treatment with respect No weapons undetected Recognizable uniforms Up-to-date equipment
IT support service (internal service provider within aircraft manufacturer)	Friendly and competent staff Hotline accessible and staffed when needed Pleasant holding music (if all lines are busy) Problems are solved (on the spot) Close-of-call e-mail with contact person (in case problems persist)

Spare-parts dispatch unit
(both external and internal
service provider)
	Friendly and competent customer contact staff
	The right spare parts at the standard quality are dispatched
	Packaging as standardized, clearly marked and clean,
	    including easy-to-understand instructions, etc., as well as
	    any other required papers

These performance objectives should be established internally first, by talking with customer contact personnel and employees who have profound knowledge of their service operation and/or experience in similar fields. In doing this, it is very important to take on the customer viewpoint to ensure that the set of objectives reflects what is assumed internally to be the customer's needs and requirements. The question 'What do we believe the customer expects from our services?' must be kept in mind and asked again and again in the process of defining these objectives.

The five performance objectives are really composites of many smaller or partial measures depending on the type and scope of the operation in question (see examples in Table 3.7). Such partial measures should now be identified for each of the performance objectives that have been internally defined. However, it is very important that only measures are chosen that are:

- easy to measure
- preferably a mix of hard and soft measures

Table 3.4   'Speed' means different things in different operations

Example service operation	What does 'speed' mean from a customer's viewpoint?
Airline (external service provider)	Fast ticket purchasing process
	Short waiting time to be checked-in
	Short time actually being checked-in
	Quick turnaround of aircraft (cleaning, food, re-fuelling, maintenance or repair if needed)
	Short waiting time to board
	Speedy onboard service (meals/drinks, etc.)
	Short waiting time to get off the plane
Airport security company (external service provider)	Short waiting times to go through security
	Short time actually being checked
IT support service (internal service provider within aircraft manufacturer)	No waiting time when calling (until someone from the support service actually picks up the phone)
	Fast resolution of problems
	Not being put on hold for long periods of time
	Not being passed around between several members of the support services
Spare-parts dispatch unit (both external and internal service provider)	Short time from receiving the order to sending the package containing the required parts
	Short travel time to the required destination

Table 3.5   'Dependability' means different things in different operations

Example service operation	What does 'dependability' mean from a customer's viewpoint?
Airline (external service provider)	The booked flight is not cancelled, nor departure delayed
	No over-booking so that some passengers have to take a different flight
	No luggage lost at destination
	Arrival at the booked destination at the scheduled time of arrival
Airport security company (external service provider)	Security check zones staffed and running when passengers arrive
IT support service (internal service provider within aircraft manufacturer)	Promises made are kept
	Low downtimes of the networks, servers and/or other IT-related systems
	The hotline is staffed and available to customers within the normal working time of users
Spare-parts dispatch unit (both external and internal service provider)	Arrival of spare parts at the required destination
	Arrival of spare parts at the required time

Table 3.6   'Flexibility' means different things in different operations

Example service operation	What does 'flexibility' mean from a customer's viewpoint?
Airline (external service provider)	*Product/service flexibility* – New technologies are considered (such as Internet services on board) and new services are offered (where economically viable) – Special requests made by individual or many customers lead to a new service offering (where economically viable) *Mix flexibility* – Wide range of inflight services offered on board (such as choice of films and audio programmes, bar, shop, massages, childcare, beauty salon. etc.) – Choice of classes (for example, first, business, economy) – Wide range of travel packages offered (flights only; more integrated offers including hotel bookings and/or car rental reservations; event bookings, etc.) – Wide choice of destinations served (either directly or via partner airlines, perhaps within a strategic alliance) *Volume flexibility* – Ability to schedule extra trips and/or deploy additional transport capacity when the need arises (such as change of type of aircraft used) – Ability to cope with higher levels of demand of individual services on board by shifting staff internally *Delivery flexibility* – Ability to cope with late arrivals and other delays due to bad weather, technical problems, etc. (for example, being able to speed up aircraft turnaround and boarding processes for the next trip; increasing cruising speed during the flight)
Airport security company (external service provider)	*Product/service flexibility* – New technologies are considered (such as registered passengers' automatic identification to accelerate security checks) and new services are offered (where economically viable) – Special requests made by individual or many customers (airport companies) lead to a new service offering (where economically viable) *Mix flexibility* – Wide range of services offered such as 'standard' security checks of passengers, luggage screening, surveillance of the airport buildings and their perimeters (including shops, parking lots and areas in the vicinity of the flight paths etc.) *Volume flexibility* – Ability to cope with very high numbers of passengers to be checked (availability of sufficient screening stations and personnel that can be deployed flexibly depending on the levels of passengers at different times during the day) *Delivery flexibility* – Deployment (places and times) of security staff flexibly adjusted to the specific airport's needs
IT support service (internal service provider within aircraft manufacturer)	*Product/service flexibility* – New technologies allowing for video conferencing and worldwide access to customers' specific IT infrastructure (servers, workstations, PCs) are considered and corresponding services offered

Example service operation	What does 'flexibility' mean from a customer's viewpoint?
	– Special requests made by individual or many customers lead to a new service offering (where economically viable)
	Mix flexibility – Wide service range offered (such as hotline services in all local customer languages or English and covering all applications/software used in the specific work environment, but also coaching and training)
	Volume flexibility – Sufficient hardware/infrastructure/licences to cope with high levels of demand – Sufficient number of staff (inside and/or outside the company) to cope with high levels of demand
	Delivery flexibility – If necessary, helpdesk members help customers face-to-face (for example, coaching) – In urgent cases the hotline remains staffed through extra hours for individual customers – Staff work extra time to meet agreed deadlines
Spare-parts dispatch unit (both external and internal service provider)	*Product/service flexibility* – New technologies and customer demand lead to new service offerings (such as automatic inflight ordering of spare parts)
	Mix flexibility – Wide range of services offered (for example, from 'standard' dispatch of spare-parts orders to complete spare-parts management solutions for individual customer airlines, etc.)
	Volume flexibility – Ability to cope with differing volumes of spare-parts orders (both in terms of the number of incoming orders and/or the variety of spare parts requested)
	Delivery flexibility – Worldwide delivery – Time from sending spare parts to delivery only limited by selected transportation type (plane, lorry, ship, train) – Delivery seven days a week around the clock

- reflect real customer concerns/needs
- can actually be influenced by the service operation considered.

Next, the selected performance objectives and their corresponding partial measures ought to be formulated in the form of individual service standards that are:

- easy to understand and as unambiguous as possible
- behaviours and actions that employees can actually influence (that is, improve or maintain)
- challenging but realistic
- accepted by employees.

Table 3.7   Some typical partial measures of performance objectives

Performance objective	Typical partial measures
Quality	Number of complaints
	Customer satisfaction scores
	Mean time between failure
	Warranty claims
	Percentage of scrap
	Number of defects per unit
	Downtime
Speed	Time from customer query to response
	Order lead time
	Frequency of delivery to customers
	Waiting time
	Total delivery time
	Customer throughput time
Dependability	Percentage of orders delayed
	Percentage of wrong deliveries
	Percentage of trainings cancelled or postponed
	Average delay of deliveries
	Perceived dependability score
	Proportion of products in inventory
Flexibility	Time needed to introduce a new product/service
	Perceived volume flexibility score
	Perceived delivery flexibility score
	Range of services
Cost	Deviations from budget
	Utilization of resources
	Cost per operation hour

Those service standards should then be discussed and refined with key customers, so that the standards selected reflect not only what are assumed to be what the customer needs and wants but what they really do need and want. Still, it is a good approach to first define those standards internally, in order to limit customer involvement to a level where the efforts necessary on their part are perceived to be justifiable. Rarely will customers be able to give a clear picture of what exactly they need, and most customers will not be motivated to spend a lot of time with researchers to find out together from scratch. Hence, if a predefined set of service standards can be presented, key customers are more likely to participate in the process of refining them until they are acceptable, because this way it simply takes less of their time. This process is summarized in Table 3.8.

Interviews and Focus Groups – Step 1

There are various ways of involving customers that need to be considered under the specific circumstances and business context of the service operation in question.

Table 3.8   Checklist – performance objectives (internal)

- Use the framework of the five performance objectives (quality, speed, dependability, flexibility and cost) to define a set of relevant and realistic performance objectives.
- Check that those performance objectives are in line with the overall strategy.
- Check that those performance objectives reflect what is perceived by front-line employees to be customer needs and requirements.
- Define partial measures for each of the objectives identified.
- Check that those partial measures are easy to measure, represent a mix of hard and soft measures, reflect assumed customer needs, and can actually be influenced by the service operation in question.
- Translate the above into a set of challenging yet realistic service standards that are easy to understand by customers and employees alike.

In the given case (that is, to help validate the proposed set of service standards), two means are particularly recommended: individual interviews and/or focus group meetings with key customers. The following points can be added to the summary in Table 3.9.

Interviews with key customers mean dealing with individual people. The advantage is that customers are less influenced by others when they make a specific statement. Disadvantages may be that interviews with the same number of candidates are likely to take longer (and therefore are potentially more expensive) than would be the case with focus group meetings, and the subjects' interaction with colleagues (other customers) cannot be observed.

Focus group meetings with key customers will involve all participants being invited to interact or discuss while giving answers or making statements. The advantages are that social interaction between customers can be observed, additional information may be revealed that would not otherwise be noticed during interviews, and the time spent with the same number of candidates is likely to be less as opposed to conducting separate individual interviews. Disadvantages would be that usually more organizational effort will be needed to prepare focus group meetings, and individual customers may feel intimidated about speaking their minds if other participants are dominant and/or promote opposed views.

Table 3.9   Checklist – interviews and focus groups

- Select the means of customer involvement (interviews, focus groups).
- Prepare a structure for interviews and/or focus groups that is based on the service standards and the selected scale of importance.

Customer Involvement – Step 1

Having developed one set of service standards that was internally produced and merely reflects what was perceived or believed to be the customers' needs and requirements, the researcher should involve key customers in this sub-step to validate and refine this set so as to obtain service standards that can serve as the service operation's guidelines and be communicated internally to every employee and externally to every customer (both existing and potential ones). Also, those customer-validated and ranked service standards can serve as benchmarks against which service performance will be measured over time, giving the opportunity for continuous improvement and better decision-making.

Customers are a valuable resource in the process of problem recognition and solution in general, not just to validate the set of service standards, insofar as they hold relevant information for problem recognition, as well as potential for both problem recognition and solution. The service operation, in turn, contributes with its technical distribution capacity, management know-how and its own potential for problem recognition and solution. By blending all the information and potentials of both the operating organization and the customer, both problem recognition and solution are significantly enhanced. In addition, customer integration is also likely to have positive side effects on long-term customer retention because it helps the company to understand its customers better and develop closer long-term relations with them.

Usually, there are too many customers to talk to them at one session. Therefore, a list of suitable key customers should be identified: this may be a mix of existing and potential customers, and should represent all major areas of service activities and customer locations served.

Out of those customers, a smaller sub-set of candidates should be selected that will actually be invited. It is important to find the happy medium between not spending too much researcher and customer time, and yet not missing important pieces of information.

Having selected the candidates to be asked for their involvement and the means by which to gain the information needed to validate the set of service standards, those customers should be invited. When doing so, the purpose of the meeting should be clearly stated and the importance of their contribution emphasized; that is, enabling the organization to improve its services to better suit their needs. No matter what means of involvement are selected, it is strongly recommended to first give the participants a clear picture of the goals of this research and why it is so important that they participate.

Only then should the set of service standards be reviewed and the scope of likely relevant comments assessed. At this stage there should be no commitment to alter a specific standard in a particular way, because later meetings with other customers may cause a change of mind and this particular customer to whom an early commitment was made may not feel taken seriously later on. There should

Table 3.10   Checklist – customer involvement (Step 1)

- Identify suitable (existing and potential) key customers and possibly select a smaller list of candidates to be asked for their involvement.
- Invite the selected candidates, explaining the purpose of the meeting and underlining the importance of their help for your ability to improve the services delivered to customers.
- During the meetings, use the set of service standards as a structure within which to retain and note all relevant inputs (comments, emotional reactions, etc.).

merely be a commitment to reviewing all comments and seriously considering all inputs given.

Table 3.10 provides a summary of the process.

Service Standards (External)

After conducting the meetings with a choice of key customers, during which all comments and/or other inputs regarding the set of initial service standards are collected, all inputs should be grouped under each service standard that was discussed so that it can be easily visualized which inputs were made by which participant concerning which service standard. Then, based on this presentation of the results from the meetings, it should be decided which of the inputs should be taken into account, because they are relevant (that is, suitable to meet customers' needs), realistic (that is, achievable), as well as meeting the criteria used to develop the initial service standards (see above).

If possible, key employees who have experience with the customers and/or know the limitations of the service operation should support the decision-making, in order not to disregard an important input to this process and to avoid defining a service standard that is unrealistic and not achievable. It should be kept in mind that, sometimes, experienced key employees can be very stuck in their old way of doing things and they may not at all be open-minded about or convinced of the usefulness of formulating service standards. This, however, should not be seen as justification not to include them in the process, because they may hold valuable information. Furthermore, those employees need to be taken on board anyway. If they are involved from early stages of the cycle, they are less likely to resist related changes later on.

Finally, the set of initial service standards should be adjusted accordingly so that a set of validated service standards (external) is created that can serve as the basis for subsequent steps of this method (the service quality cycle).

This part of the process described above is summarized in Table 3.11.

Having described in this section the approach of generating a set of internally and externally validated service standards, the next section describes how these

Table 3.11   Checklist – service standards (external)

- Collect all comments and other inputs made by the participants during the interviews and/or focus group meetings.
- Group those per each service standard that was discussed.
- Decide which of those inputs are justified and can or should be taken into account.
- If possible, use key employees to help you make that choice.
- Adjust the set of service standards accordingly.

---

standards can be ranked by key customers so that the researching organization can put in visual form the relative importance customers attach to each of them.

Step 2 – Ranking of Service Standards

The second step (see Figure 3.4) is about having the service standards that were generated and validated in the previous step ranked by the customer so that it becomes clear how much importance customers attach to each service standard defined. To know this is absolutely essential, because it helps the organization to decide where to focus its efforts and how to use employees and/or allocate resources. The analysis of performance measurements later in the service quality cycle is greatly enhanced by

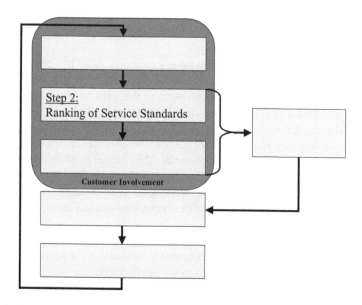

Figure 3.4   Ranking of service standards

Anticipating Recovery Opportunities?

Airline operations
Examples for recovery opportunities are delayed or cancelled flights, over-booked flights leading to individual customers having to wait for another flight, missed connecting flights, failure to serve pre-ordered special food on board, blocked toilets, drinks spilled over customers and many others. All these incidents bear the potential for bitter disappointment of customers, resulting in them ceasing to use the airline and, even worse, very bad word-of-mouth communications, with potentially disastrous consequences for future sales. Therefore, all such incidents (along with their causes, steps taken and feedback) need to be recorded; best practice approaches to similar incidents must be established and grafted into the training process to improve future reactions and effectively turn around angry customers.

Airport operations
Massive queues at check-in and security checkpoints, flight delays due to lack of ground support resources, strikes, bomb alerts, fires, as well as flight delays due to bad weather with the resulting numbers of waiting passengers in the airport are all examples of recovery opportunities. Airport operations need to be carefully prepared for those in order to minimize the impact on customer satisfaction. This is also likely to cover contractual aspects with security sub-contractors as well as agreements and plans with the local fire and police departments for the worst cases. Pre-planned crisis management must be able to be triggered at short notice, whereas customer flows in the airport (such as arrival, check-in, security, boarding) ought to be managed as part of the ongoing operations.

Aircraft manufacturing operations
One of the most critical examples for a recovery opportunity is a customer airline's 'aircraft on ground' (AOG) because this presumably causes the airline a lot of trouble. It is not just that the aircraft does not earn money whilst grounded but also that passengers will have to wait and/or accept changes in the airline's planned flight schedule. It is of utmost importance that aircraft manufacturers have best practice approaches prepared and staff trained to solve such situations as quickly as possible.

putting performance against the declared service standards in relation to the relative importance of the latter, as perceived by the customer.

Step 2, the ranking of service standards, should be taken once during the implementation of the service quality cycle and thereafter reviewed periodically, depending on the volatility of customer perceptions of the importance of individual service standards and/or changes in the situational context that may require strategic adaptation of the service operation over time (see Step 6).

As mentioned before, it is critical to find the right balance of customer involvement, not annoying customers with constant requests (questionnaires, interviews, etc.)

but still getting enough vital information to be able to put the service operation in line with customer needs. It may be a good approach to combine the actual customer involvement events from Steps 1 and 2; that is, asking key customers to rank the service standards directly after they have given their inputs concerning each individual service standard. Although individual key customers would only be bothered with one meeting for Steps 1 and 2, this comes with the risk that the service standards presented may still change significantly and so may the perceived importance of some of the standards.

In order to obtain a set of customer-ranked service standards, a number of sub-steps are proposed here (see Figure 3.5).

The starting point of Step 2 is the set of customer-validated service standards produced in Step 1. A decision needs to be taken on the scale of importance that will be used to rank the standards. Next, the means of involving the customer should be selected and prepared (interviews, questionnaires, focus groups), followed by the actual customer involvement. The outcome of this step is a suitable table of ranking results that will be used during Step 4.

Service Standards (External)

The validated set of customer-defined service standards as generated in Step 1 serves as the starting point and basis of this step. This set of service standards reflects

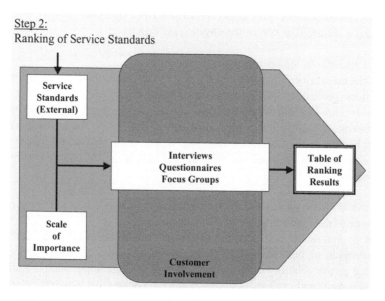

Figure 3.5   Step 2 – Ranking of service standards

Table 3.12   Checklist – scale of importance

- Select a suitable scale of importance.
- Check the scale is sufficiently detailed to allow for clear ranking of service standards, yet immediately understandable for customers.

what customers really want the service operation to deliver, although the relative importance of each standard to customers is not known yet.

Scale of Importance

The sub-step called 'scale of importance' consists of selecting a suitable scale to measure the relative importance of each service standard as perceived by customers. In principle, any scale that allows for making the rough distinction between very important, important and less important standards will be useable; for example, a scale from 1 (not important) to 10 (very important). However, in order to get more accurate feedback from customers, it is important that each value is clearly defined and everybody immediately understands this definition. The nine-point scale of importance as discussed in Chapter 2 (Table 2.3) is an example of a well-defined scale that is intuitive yet sufficiently detailed to allow for clear ranking of service standards.

Table 3.12 summarizes the above points.

Interviews, Focus Groups and Questionnaires – Step 2

There are three main approaches recommended here to collect systematic customer feedback: conducting interviews with key customers, sending out questionnaires to all or specific customers, and conducting focus group meetings with key customers. All three approaches should be considered in light of the specific context of the service operation in question before one or several means are selected. The choice of means also depends on whether or not the customer involvement events of this step are combined with those of the previous step. If they were, customers would be asked during the interviews and/or focus group meetings not only to comment on the initial set of service standards (part of Step 1) but also to rank them (part of Step 2), with the advantages and risks mentioned above. Table 3.13 summarizes the process, but there are some additional points, as made below.

For *interviews* and *focus groups*, refer back to the descriptions in Step 1. *Questionnaires* represent the least personal form of customer contact in feedback collection. Advantages of using questionnaires are that:

- many customers can be reached in a very short time and at least give the opportunity to give feedback (although many will not make use of it);
- all customers are asked the same questions;
- collection and use of the reply forms for later analysis is straightforward;

- if the questionnaires used are based on the service standards selected (that is, if the standards are clearly stated in the questionnaire, which is recommended), this represents a good opportunity to communicate the service operation's standards to customers.

Disadvantages include that the reply rate is usually very low – most customers do not even take the few minutes needed to fill in the form. Also, the questions are likely to be understood differently by different customers; in other words, the service standards may not be properly understood if they are not well formulated. Unlike in personal interviews, where you may feel the need to investigate further in a specific direction and can ask more detailed questions, questionnaires do not offer this opportunity because there is no direct interaction. The need to investigate further may not even become apparent if the customer does not make corresponding comments in the reply form.

If located in a non-English-speaking country, questionnaires should be prepared in the local language but also in English where appropriate, so that foreign customers do not feel excluded and, even more importantly, are able to give their feedback too. Both versions should be sent out together to give everybody the choice in which language he/she wants to reply.

Customer Involvement – Step 2

If interviews and/or focus group meetings are used, the same comments apply as detailed for Step 1 in terms of:

- identifying (existing and potential) key customers
- selecting and inviting participants from that list
- specifying the purpose
- emphasizing the value of their contribution, such as 'enabling us to improve our services to better suit your needs'.

Table 3.13   Checklist – interviews, focus groups and questionnaires

---

- Select the means of customer involvement (interviews, questionnaires, focus groups).
- If interviews and/or focus groups are used, prepare a structure for those (based on the service standards and the selected scale of importance).
- If a questionnaire is used (based on the service standards and the selected scale of importance), formulate it carefully and as clearly as possible to limit the room for different interpretations.
- Where appropriate, prepare two versions of the questionnaire, one in the local language and one in English.
- Check the questionnaire is clear, short and simple.

No matter what means of face-to-face involvement are selected, it is strongly recommended to first give the participants a clear picture of what you are trying to do and why it is so important that they participate. Only then should you go through the set of service standards and take in all relevant comments you may get. Asked which service standards are the most important ones to them, they are likely to answer that all are equally important to them and no one must be neglected. It should then be emphasized that there may be shortages of resources at a given point in time to provide the services offered and decisions will have to be taken on the basis of such a ranking of service standards. Therefore, it is most important to have a clear view on the relative importance of each standard to customers, so that decisions can be taken in their best interests.

If questionnaires are used, the purpose of the questionnaire should be stated and the importance of each customer's participation underlined in the covering letter or e-mail. If it is possible during ongoing service encounters with customers, they ought to be told beforehand that a questionnaire is about to be sent to them, and that it would be very important if they could give their feedback and send it back in order to improve the services offered. This approach has proven to increase the rate of questionnaire replies because there is anticipated compensation for the lack of personal contact when receiving a questionnaire and many customers will feel more obliged to reply if they have spoken about it with somebody they know from the service operation.

Table 3.14 gives a summary of the process just described.

Table of Ranking Results

After conducting the meetings with a choice of key customers, during which all comments and/or other inputs regarding the set of initial service standards are collected, and/or when all replies to a questionnaire have been received, all inputs should be grouped by each service standard and possibly by customer team or segment (but only if there are significant differences in their ranking behaviour). By doing this it can be easily envisaged how each service standard was ranked by which customer team or segment or overall, as well as which additional inputs were made by which participant concerning which service standard.

Finally, the ranking results should be presented in a table or spreadsheet, which can be very simple if all customers showed similar ranking behaviour across all service standards, slightly more detailed if specific customer teams or segments showed different ranking behaviour. If there are no significant differences in the perceived importance of each service standard across all customers, the average ranking of the service standards will be sufficient.

Table 3.15 provides a summary of this part of the process.

In this step, having identified and put in visual form what relative importance customers attach to each of the service standards generated during Step 1, the following step covers performance measurements against those service standards.

Table 3.14   Checklist – customer involvement (Step 2)

- If interviews and/or focus groups are used, identify suitable (existing and potential) key customers and possibly select a smaller list of candidates to be asked for their involvement.
- Invite the selected candidates, explaining the purpose of the meeting and underlining the importance of their help for your ability to improve services delivered to customers.
- During the meetings, use the set of service standards as a structure within which to retain and note all relevant inputs (comments, emotional reactions, etc.).
- If questionnaires are used, send them out to all or specific customers.
- Tell customers you meet during normal service encounters that a short questionnaire is going to be sent out to them and that it would be very helpful if they could participate (this is likely to increase the reply rate).

Step 3 – Measuring Performance against Service Standards

The third step (see Figure 3.6) is about measuring the service operation's performance as perceived by customers against the service standards that were generated, validated and ranked in the previous steps. To do this is absolutely essential for the following reasons:

- First, the operation's performance is being measured as real customers perceive it, and if they do not believe that better services are being offered than those of competitors, they are not likely to use or keep using the original service provider.
- Second, the criteria for these measurements are the very service standards previously defined and also communicated to customers. In other words, the organization is measuring what it is really trying to achieve with its service operation and what customers really expect from it.

The analysis of performance measurements in the next step of the service quality cycle (Step 4) is greatly enhanced by putting customer-perceived performance

Table 3.15   Checklist – table of ranking results

- Collect all replies and other inputs made by participating customers.
- Group them per service standard.
- Present the ranking choices in a spreadsheet (table).
- If there are big differences between specific customers' rankings, highlight those differences.
- If almost all customers show the same ranking behaviour, use the average ranking of the service standards.

Involving the Right Customers?

Airline operations
Airlines have a variety of customers, the most important ones of whom can be placed into different customer segments such as business travellers, holidaymakers, weekend commuters and others. It is important to involve customers from all major segments to get the feedback and information needed in order to effectively improve perceived service quality. One good approach is to hand out questionnaires during a flight and offer little presents for participants or give a bottle of champagne to one of the participants before landing. Another approach could be to invite a number of passengers who are waiting for their flight to a restaurant in the airport or the airline's lounge to hold a focus group meeting with them or to conduct individual interviews.

Airport operations
Airports have a number of different types of customers (such as airlines and their passengers) using the airport, shops and restaurants situated in the airport buildings, plus car rental companies operating from the airport, and others. It is essential to involve representative or key customers from all these types of customers in order to get the needed information.

Aircraft manufacturing operations
Aircraft manufacturers' customers are predominantly airlines (civilian aircraft) and governmental institutions (mainly military aircraft). Internally, they represent a complex network of internal customers and suppliers across all business areas and sites. Enabling services or support services, for instance, are delivered to development and design teams which, in turn, deliver services to actual aircraft manufacturing teams. The latter deliver complete aircraft and necessary support services to delivery teams. But also the finance department delivers services to general management and human resources teams deliver their services to engineering teams by providing them with the right (highly qualified) resources to join those teams. No matter who are the external and/or internal customers of a service provider, key customers from all types of customers served should be involved.

against the declared service standards, and in relation to the relative importance of those as ranked by customers.

Step 3 (that is, the measurement of performance against the service standards) should be taken regularly. During the implementation of the service quality cycle those measurements will be needed more frequently. Once the cycle is well established, measurements should be taken less frequently, depending on the volatility of customer expectations and their perceptions of the importance of individual service standards, and/or changes in the situational (business) context that may require strategic adaptation of the service operation over time (see Step 6).

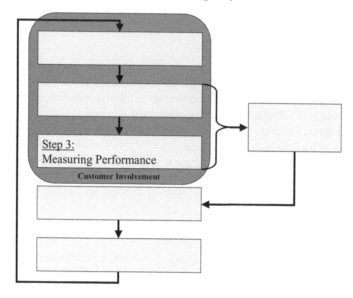

Figure 3.6   Measuring performance against service standards

As mentioned before, it is critical to find the right balance of customer involvement, not annoying customers with constant requests (questionnaires, interviews, etc.) but still getting enough vital information to be able to put the service operation in line with customer needs.

In order to obtain a meaningful set of performance measurements (per service standard), a number of sub-steps are proposed here (see Figure 3.7).

The starting point of Step 3 is the set of customer-validated service standards produced in Step 1. A decision needs to be taken on the scale of performance that will be used to rate the service operation's performance against these standards. Next, the means of involving the customer should be selected and prepared (interviews, questionnaires, focus groups), followed by the actual customer involvement. The outcome of this step is a suitable table of performance results that will be used during Step 4.

Service Standards (External)

The validated set of customer-defined service standards as generated in Step 1 serves as the starting point and basis of this step. This set of service standards reflects what customers really want the service operation to deliver.

Step 3:
Measuring Performance against Service Standards

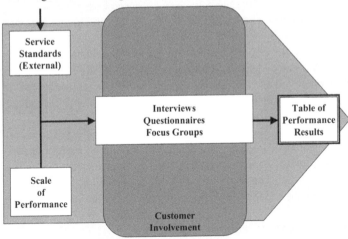

Figure 3.7   Step 3 – Measuring performance

Scale of Performance

The same comments apply as made for Step 2 regarding selection of a scale of performance and obtaining accurate feedback from customers. The nine-point scale of performance as discussed in Chapter 2 (Table 2.4) is an example for a well-defined scale that is intuitive yet sufficiently detailed to allow for clear rating of performance against service standards.

Types of Service Operation

There is a distinction to be made between two basic types of service operations that have direct influence on the type of the scale of performance to be selected.

In *Case 1*, the operation under consideration is serving both *external customers* (against direct competition in the market) and *internal customers*, but where those internal customers actually possess the freedom to choose their service provider (for example, if they are organized as profit centres and can either outsource the operation or have it performed by an internal supplier).

In *Case 2*, the operation considered is serving *internal customers* and those customers cannot choose to use an external supplier (for example, if they are organized as cost centres and have to make available part of their allocated budgets for the services provided). They may, however, be in the position to either reduce the budget spent on the services provided or cut those services out altogether if they are not convinced it is worthwhile to 'pay' budget for those services.

In either of the above cases it is essential to measure performance and to achieve at least a certain degree of customer satisfaction along costumer-defined service standards. The difference between Case 1 and Case 2 lies in the approach to measuring performance and to present it visually it in the importance-performance matrix later on.

Case 1 allows for customer-perceived performance measurements against the perceived performance of direct competitors as suggested in Chapter 2 (see Table 2.4). The service operation's performance as perceived by the customer can be grouped into 'better than', 'same as' and 'worse than' competitors, with each category being divided into the three sub-groups 'strong', 'medium' and 'weak'.

Case 2 requires a slightly different approach, because there is no direct competitor with which customers could compare the service operation's performance. Customer-perceived performance measurements must be taken against the defined service standards, for instance on a scale from 1 ('I do not think at all that the specific service standard has been met') to 10 ('I fully agree that the specific service standard has been met').

Table 3.16 summarizes the above points.

Interviews, Focus Groups and Questionnaires

These three approaches to collecting systematic customer feedback have already been detailed in Steps 1 and 2 and summarized in Table 3.13. In any given case, all three approaches should be considered in light of the specific context of the service operation in question and one or several means selected.

Customer Involvement – Step 3

As noted for Steps 1 and 2, using interviews and/or focus group meetings involves:

* identifying key customers
* selecting and inviting participants from that list
* specifying the purpose

Table 3.16   Checklist – scale of performance

---

* Depending on the circumstances, decide whether the scale of performance should either reflect performance as compared to competitors (profit centre, own business, etc.) or very poor versus very good performance (cost centre, internal service operation, etc.).
* Select a suitable scale of performance.
* Check that the scale is sufficiently detailed to allow for meaningful performance measurements against the declared service standards.
* Check that the scale is immediately understandable for customers.

- emphasizing the value of their contribution for improvement of the service offered.

Whatever means of face-to-face involvement are selected, it is strongly recommended to first give the participants a clear picture of the goal of the exercise and why it is so important that they participate. Only then should the set of service standards be gone through and all relevant comments taken in.

There are two types of information that should be gained for each service standard. First, there should be a clear value or standardized statement; for example, '1 – Considerably better than competitors' (if the scale of performance from Chapter 2 is used): the response to this type of statement is described as quantitative information. Second, any additional background information or feedback from the customer should be taken note of, because it will help to put the performance values into perspective and gain a more differentiated view on why customers are driven to come up with a specific value: this is described as qualitative information.

The value and use of questionnaires has been noted under Step 2 (see 'Customer Involvement – Step 2'. The same advantages and disadvantages apply for Step 3.

The checklist in Table 3.17 should be used at this step.

Table of Performance Results

Having gathered all the input on performance from interviews, focus groups and/or questionnaires, the results should be grouped by service standard. They could also possibly be grouped per customer team or segment if there are significant differences

Table 3.17   Checklist – customer involvement (Step 3)

- If interviews and/or focus groups are used, identify suitable (existing and potential) key customers and possibly select a smaller list of candidates to be asked for their involvement.
- Invite the selected candidates, explaining the purpose of the meeting and underlining the importance of their help for our ability to improve our services delivered to them.
- During the meetings use the set of service standards as a structure within which to retain and note all relevant inputs (comments, emotional reactions, etc.).
- Take note of both types of information, not only the measurement values per service standard, but also any other background information as to why specific customers came up with a given value.
- If questionnaires are used, send them out to all or specific customers.
- Tell customers you meet during normal service encounters that a short questionnaire is going to be sent out to them and that it would be very helpful if they could participate (this is likely to increase the reply rate).

Table 3.18   Checklist – table of performance results

- Collect all replies and other inputs made by participating customers.
- Group them per service standard.
- Present the performance measurements in a spreadsheet (table).
- If there are big differences between specific customers' perceptions, highlight those differences.

---

in their rating behaviour, so that it can be easily grasped how the performance per service standard was rated by which customer team or segment or overall, as well as which additional inputs were made by which participant concerning which service standard.

Finally, the performance measurement results should be presented in a table or spreadsheet, which can be very simple if all customers show similar rating behaviour across all individual service standards, slightly more detailed if specific customer teams or segments perceive performance very differently. If there are no significant differences in the perceived performance against each service standard across all customers, the average value will be sufficient.

Table 3.18 sums up this part of the process.

Having, in this step, measured and visually established how customers perceive the service operation to perform against the service standards that were previously defined (Step 1) and ranked (Step 2), the following step covers the analysis of the current situation which is partly based on those performance measurements.

Step 4 – Analysing the Current Situation

The fourth step (see Figure 3.8) is about analysing the current situation of the service operation in terms of customer-perceived performance against service standards in a given situational context.

Both operational and strategic decision-making is greatly enhanced by taking into account what matters to customers (service standards) and how they perceive the service operation to perform (against those service standards), but also what is the situational context of the service operation. In other words, a thorough analysis of the situation at a given point in time is highly likely to result in better decisions. The purpose of this step is to systematically create such a basis for decision-making.

Step 4 is the analysis of the current situation based on performance measurement results, which should be taken regularly, at least every time performance is measured but also when changes in the situational context occur that may have a significant impact on the service operation and/or its customers. The analysis resulting from this step will serve as the basis for operational and strategic decision-making in Steps 5 and 6.

In order to analyse the current situation of the service operation a number of sub-steps are proposed here (see Figure 3.9).

Which Scale of Performance?

Airline operations
Performance is likely to be measured by asking passengers to rate the airline's performance against their declared service standards. Since customers will (more often than not) be comparing the airline's services with those offered by competing airlines it may be a good idea to decide on a scale of performance that reflects performance against competitors. However, some passengers who do not fly regularly will not feel they have a good basis to make such comparisons and therefore may not be willing to participate.

Airport operations
Passengers, airlines, car rental companies and other customers can and usually do compare the airport operation's performance with other competing airports. Hence, the chosen scale of performance could either reflect performance against competitors or very good versus very bad performance. The former allows for direct yet not very specific comparisons with competitors. If, on the other hand, most individual shops or restaurants only exist in this one airport, the scale of performance should reflect very good versus very bad performance because those customers presumably are in a worse position to actually compare the airport's service quality with that of other airports.

Aircraft manufacturing operations
External customers are very much more likely to be asked to compare the services offered with those of other aircraft manufacturers; for instance, between the big rivals Airbus and Boeing. The reason for this is that most customer airlines or governments have been able to gain experience in using services from several aircraft manufacturers. Furthermore, since there are not many competitors in this market, it will be of high interest to aircraft manufacturers to be specifically measured against the performance of those. Internally, the selected scales of performance are likely to reflect very good versus very bad performance because there is usually no direct competition between internal service providers and other (also external) service providers offering the same spectrum of services.

The starting points of Step 4 are the tables of the performance results (prepared in Step 3) and the ranking results (prepared in Step 2). Both tables will be used to display in visual format – in the form of an importance-performance matrix – the service operation's performance per service standard at a given point in time.

If in Steps 2 and/or 3 significant differences between groups of customers or segments were detected, these differences can (and in fact should) be taken into account by producing several importance-performance matrices. For instance, if one customer team perceives particular standards (out of the same set of service standards) to be more important than all other customers, this should be presented

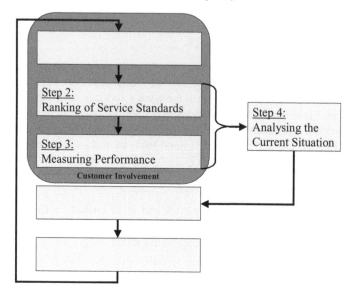

Figure 3.8   Analysing the current situation

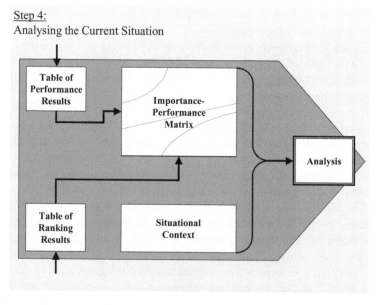

Figure 3.9   Step 4 – Analysing the current situation

in visual form by producing an extra importance-performance matrix just for this team.

Based on this visualization and consideration of the relevant situational context, the current situation per service standard can be analysed. The outcome of this step is a thorough analysis per service standard that will allow for the derivation of concrete action items in Step 5.

Table of Performance Results

The table of performance results that was produced during Step 3 is one of the starting points of the current step. This table shows on a per service standard basis how customers (overall, or individual segments or teams if there were significant differences in measurements) have rated the service operation's performance.

Table of Ranking Results

The second starting point of this step is the table of ranking results which was produced during Step 2 and which shows what relative importance customers (overall, or individual segments or teams if there were significant differences) attached to each individual service standard; that is, which standards were perceived to be 'order winners', 'qualifiers' or 'less important'.

Importance-Performance Matrix

Based on the tables of performance and ranking results that were produced in Steps 2 and 3, the service operation's performance can now be put into the form of an importance-performance matrix that shows the customer-perceived performance against individual service standards, depending on the customer-perceived importance of the service standard in question.

First, a generic importance-performance matrix should be drawn (also see Chapter 2, Figure 2.15), showing the scales of importance and performance (that were selected in Steps 2 and 3) on the x and y axes respectively.

The areas of 'excess' and 'urgent action' need to be positioned in the left upper corner and the right lower corner respectively, as shown in Figure 3.10. The 'lower bound of acceptability' should be a straight line from the lower third of the left limiting line of the matrix to about two thirds of the height of the right limiting line.

It should be kept in mind that the matrix serves as a simple yet effective and intuitive tool to visualize the position of service standards, based on their perceived importance and the service provider's customer-perceived performance against those standards. However, the exact position in the matrix that results from measurements must not be overvalued, especially if the number of participating customers was low and/or a combination of several approaches to involve customers (such as questionnaires and interviews) was used. The position in the matrix merely gives

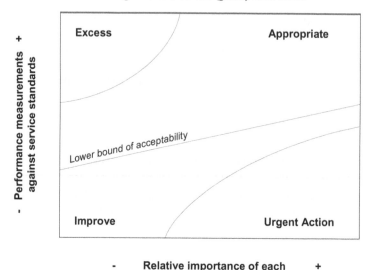

Figure 3.10   Generic importance-performance matrix

the service provider a rough (yet sufficient) indication of where it stands in the eyes of customers.

Once this generic importance-performance matrix is prepared, each service standard should now be allocated an easily recognizable symbol; for instance a small circle with the number of that service standard. Then, for each standard, the selected symbol should be positioned in the matrix according to the ranking (x axis) and performance measurement results (y axis) from Steps 2 and 3 respectively. Table 3.19 summarizes this process.

Situational Context

After going through the previous sub-step, a 'snapshot' will be apparent of what relative importance customers attached to each individual service standard and how the service operation was perceived to perform against each service standard.

In order to understand and interpret this snapshot, however, the situational context must be considered. It has been shown that the ranking of individual service standards does not tend to be as volatile over time as the performance measurements usually are. Therefore, it will (more often than not) be essential to be able to detect causal relationships between a specific performance measurement and internal or external factors that may have influenced customer perceptions of performance against the service standard in question.

First, the overall situational context should be considered in that, although it cannot directly be influenced by the service operation nor its customers, it may

Table 3.19   Checklist – importance-performance matrix

- Draw a generic importance-performance matrix.
- Use the scale of importance that was selected during Step 2 on the x axis.
- Use the scale of performance that was selected during Step 3 on the y axis.
- Select a recognizable symbol for each standard in the matrix (such as a small circle with the number of the service standard from the whole set of selected standards).
- Look at each service standard individually and insert the corresponding symbol into the matrix, using the ranking value on the x axis and the rating value on the y axis.

---

still help to explain certain factors and behaviours. For instance, if the climate in a customer organization is very tense because employees have been made redundant and/or time pressure is tremendous, those factors may not be directly influenced by individual customers but are still likely to contribute to their perceptions of the importance of individual service standards and performance against them.

Second, internal factors that can be directly influenced by the service operation itself (or its employees) should be considered. For instance, if the human resources to fulfil a specific task were changed and somebody new took over a specific customer contact activity, this change may well have influenced customer perceptions of the services delivered.

Third, external factors that can be directly influenced by customers should be considered. For example, if a customer's management has decided to cut the budget allocated to an internal service provider, the latter will have to adjust their service offerings to the new financial situation, perhaps reducing certain aspects of the services delivered. This, in turn, is likely to affect individual customers' perceptions of the services delivered.

Table 3.20 gives a checklist of these points.

Analysis

The analysis should be in written form and start with a brief description of the overall situational context that cannot be directly influenced by the service operation, or by its customers. Then, for each individual service standard, the following questions should be answered.

Where does it stand in the matrix and what does this mean? For instance, a specific service standard for a traditional airline such as 'friendliness of flight attendants' may be positioned in the 'urgent action' zone. This means that it is considered by customers to be important but the service operation's performance against this standard is perceived to be poor.

What internal reasons may there be for the current position? In the given example, internal reasons that could be influenced by the service operation (the airline) may be that flight attendants have to work longer hours and receive less compensation

Table 3.20   Checklist – situational context

• Consider the overall situational context of the service operation and its customers that has not directly been influenced by either the service operation or its customers.
• Consider internal factors that have been influenced by the service operation; that is, its employees (such as recent changes in the scope of services or how they are delivered to customers, concrete actions to improve customer-perceived performance, etc.).
• Consider external factors that have been influenced by customers (such as any recent changes in using the services offered).

payments for long-haul trips because the company has just implemented a cost-cutting programme and made several flight attendants redundant. Most flight attendants may feel insecure about their own job and frustrated about receiving less money for more work. This may lead to bad motivation amongst them and cause customers to perceive cabin staff as unfriendly.

What external reasons may there be for the current position? External reasons for the current position could be that a company had usually sent its employees (individual customers) to travel with that airline on business trips and the company may have changed its policy on business trips insofar as its employees are no longer allowed to fly business class but now have to fly economy class, for cost-saving reasons. Those customers from this group that have given feedback to the airline might have rated performance on 'friendliness of flight attendants' lower because they are not used to flying economy class and subconsciously perceive the airline to perform worse against that service standard.

Is that position acceptable under the circumstances? If not, where should the new position in the matrix be? In the given example, the position of the service standard in the importance-performance matrix is definitely not acceptable. As the name of the zone suggests, urgent action is needed to return to the 'appropriate' zone, if customers are to be retained and negative word-of-mouth communications are to be avoided or at least limited.

If not, what alternative options are there and what are their advantages and drawbacks? The alternatives to cope with the external factors are very limited in the given example, because the service operation (the airline) probably cannot influence the customer company's decision-making on its business travel policy. Also, and even more importantly, 'friendliness of flight attendants' is essential, especially for traditional airlines serving long-haul routes. So you cannot easily influence customer perceptions in a way so that they perceive friendliness as less important if they fly economy class. This could be done with hot meals on board, champagne, newspapers, etc., but not with the very basis of dealing with customers – being friendly and respectful.

However, there are likely to be alternatives to cope with the internal factors; that is, the motivation problems of the flight attendants have to be tackled, so that customers perceive them as friendly again. Some options should be considered that might be useful – individually or combined with other options – to move the service standard from its current position in the matrix to a new improved position. Consider the possibilities listed below.

Do nothing is always an option that ought to be considered. The advantages are that there are no changes, no additional resistance to deal with, no extra costs and insecurities about whether the new situation will be better or not. A major drawback, in turn, would be that if there is a problem, which is the case in the given example, the problem remains. It is likely that customers would eventually turn away from that airline and – all other conditions being equal – choose a different airline. Worse, they would be likely to spread negative word-of-mouth opinions about the airline in question to other existing or potential customers. To summarize, this option could not be recommended in the given case but still belongs in the analysis to give decision-makers the full picture and let them decide on the basis of a well-balanced analysis.

Increase awareness of the importance of friendliness amongst the airline's flight attendants. Organizing relevant seminars or coaching sessions with all flight attendants of the airline could achieve this. Advantages would be that cabin staff might feel invested in, and they would probably change their attitude out of personal conviction that friendliness is important. Drawbacks would be that these measures cost extra resources (working time of cabin staff and seminar or coaching fees), and the underlying causes of the current lack of motivation are not addressed.

Offer incentives for friendliness towards customers. Advantages would be that positive behaviour is not only promoted but it actually pays for cabin staff to be friendlier to customers. Drawbacks would be that the measurement of individual friendliness is not easy and could cause conflicts and frictions between members of cabin staff. Also, such a scheme would cost money and time to implement. The underlying causes of the current lack of motivation would still not be addressed.

Remove some or all of the initial reasons for bad motivation amongst cabin staff; that is, turn back some or all of the decisions taken by the airline management such as the increase in workload and reduction of compensation payments. Advantages

Table 3.21   Checklist – analysis

• Briefly describe the overall situational context that cannot be directly influenced by the service operation, nor by its customers.
• Interpret each service standard individually.
• Where does it stand in the matrix and what does this mean?
• What internal reasons might there be for the current position?
• What external reasons might there be for the current position?
• Is that position acceptable under the circumstances (given context)?
• If not, where would be an acceptable position in the matrix?
• What are the alternative options to shift its position in the matrix?
• What are the advantages and drawbacks of each option?

Analysing the Current Position?

Airline operations
One of a low-cost airline's service standards is to be very punctual. Customers flying from a specific airport rate performance to be slightly worse than that of a traditional airline and rank punctuality as a qualifier of medium importance. This particular measurement lies just in the improve zone of the matrix. The main reason seems to be lack of sufficient ground support equipment at the small airport that needs to cope with up to three aircraft simultaneously (external reason). The position is not acceptable and should be moved up into the appropriate zone in the medium run. One option to achieve this would be to put pressure on the airport services provider to adjust their infrastructure.

Airport operations
One of an airport's service standards is to keep passengers' waiting time at security checkpoints as short as possible. Customers rate performance to be slightly worse than that of competing airports and rank this service standard to be very important. The position lies in the urgent action zone of the matrix and is not acceptable. Immediate action is needed to shift it up into the appropriate zone. One option to improve the situation would be to increase the number of security checkpoints in the airport.

Aircraft manufacturing operations
One of the service standards of a specific engineering department is 24-hour availability of one of their experts for a mobile maintenance crew. Customers (members of this crew) rank this standard as the highest priority but rate performance as below medium only. The position in the matrix lies in the urgent action zone and must be changed immediately. Reasons may be a recent change of the entire phone system of the company (external) and/or lack of awareness of the importance of round-the-clock availability amongst the department's experts on duty (internal). The position is not acceptable and must be shifted towards the right upper corner of the matrix urgently. One option to achieve this would be to increase awareness by inviting customers to a meeting with the experts on duty, during which customers are given the opportunity to explain why it is so important to them.

would be that cabin staff would feel more motivated, which is likely to result in friendlier behaviour towards customers. Drawbacks would be that the financial problems of the airline that had led to these drastic decisions of sacking employees, cutting compensation payments and increasing the work load for cabin staff, are still not resolved. Also, those managers that made these decisions may think they lose face inside the airline, because they would basically be saying 'we were wrong and we take back our decisions'. On the other hand, it could well be argued that managers actually show strength if they take back bad decisions.

Table 3.21 presents a checklist of the points made above.

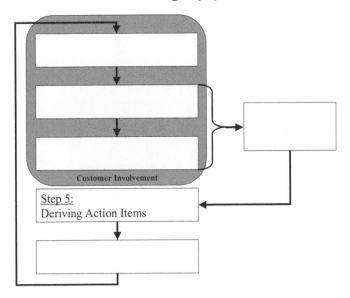

Figure 3.11 Deriving action items

Having analysed, in this step, the current situation of the service operation, taking into account customer-perceived performance against service standards as well as the situational context of the service operation and its customers, the next step consists of deriving concrete action items from this analysis.

Step 5 – Deriving Action Items

The fifth step (see Figure 3.11) is about deriving concrete action items from the analysis produced in the previous step. Those action items must be based on the full picture that is represented by the analysis, including both performance measurements and the situational context, in order to effectively improve customer-perceived service quality over time.

Whilst Step 4 (the analysis) culminates in the description of options for possible action items per service standard, also covering a critical discussion of the pros and cons of each option, this step is about actually making a choice of what exactly should be done concerning each service standard in order to improve perceived service quality.

Step 5 (the derivation of action items based on the previous analysis) should be taken regularly, at least every time an analysis is carried out. In order to derive concrete action items per service standards a number of sub-steps are proposed here (see Figure 3.12).

Step 5:
Deriving Action Items

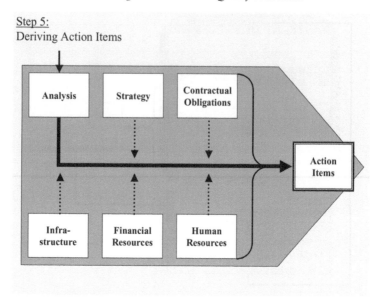

Figure 3.12   Step 5 – Deriving action items

The starting point of Step 5 is the analysis produced during Step 4. The purpose of this step is to derive from this analysis and from a number of additional considerations (see below), a set of practical action items: these represent individual operational and/or strategic decisions that are taken in order to improve the service operation's service quality as perceived by customers. Those additional considerations needed during the process of deriving action items from the analysis are described in the sub-steps of *strategy, contractual obligations, infrastructure, financial resources* and *human resources*. The outcome of this step will serve as the starting point for the next step, 'closing the cycle', which will cover considerations of how and when the action items can be best implemented and followed up over time.

Analysis

The analysis, being the result of the previous step, forms the basis from which action items can be derived in a systematic manner. It offers a short but balanced discussion of each service standard, its position in the importance-performance matrix and the situational context that is likely to have influenced the current position, as well as a number of options how the position could be maintained (if it is satisfactory) or improved (if it is not satisfactory). However, in order to come up with practical action items, some additional considerations are necessary.

Table 3.22   Checklist – strategy

- What are the overall short-, medium- and long-term strategic goals of the customer organizations?
- What are the overall short-, medium- and long-term strategic goals of the service operation?
- What is the expanded marketing mix of the service operation (product/ service, place, promotion, price, people, physical evidence, process)?
- Going through all service standards one by one, which of the options presented in the analysis are in line with the customers' and the service operation's strategic goals and fit the current expanded marketing mix of the service operation or could be adjusted to fit?

Strategy

Before any action items are decided upon to improve service quality, it should be considered whether or not they fit in the strategy of both the service operation and the customer organization(s). Therefore, it is essential to find out (where possible) the short-, medium- and long-term strategic goals of both the service operation and its customer organization(s). The whole set of action items will have to fit into the full picture and support those strategic goals, if any improvements are to endure and long-term customer relationships are to be built and maintained.

This also means that the service operation's strategic positioning (how the customers view or should view in the future the service operation to stand out in the market), as well as the expanded marketing mix (product/service, place, promotion, price, people, physical evidence, process – see Table 2.1) that is used to reach that position in customers' minds, must be considered, in order to ensure that the set of action items is in line with the strategic positioning and consistent with the marketing mix used.

With the above strategy and marketing considerations in mind, the options that were presented in the analysis should be screened one by one to find out which of them do actually fit or could be adjusted to fit.

Table 3.22 summarizes the points of this aspect.

Contractual Obligations

Next, any contractual obligations that exist between the service operation and its customers should be taken into account, such as service letter agreements (for internal service providers) or actual contracts (for external service providers). It may well be that certain actions are not possible because they would represent a non-compliance with such a (legally) binding document. Based on this consideration, the options that were presented in the analysis should be screened one by one to find out which of them are ruled out by those contractual obligations. Table 3.23 presents a checklist which could be used at this stage.

Table 3.23   Checklist – contractual obligations

• What are the current contractual obligations the service operation has to fulfil?
• Going through all service standards one by one, which of the options presented in the analysis are ruled out by those contractual obligations?

Infrastructure

Another limitation to action items may be the current infrastructure, such as hardware (servers, PCs, laptops, mobile telephones, etc.), software tools, licences, office space, as well as means of transportation to actually go to customers. Therefore, it is important that the current infrastructure available to the service operation is considered and it is checked whether any lack of infrastructure could be overcome in the near future, and how this could be achieved. Based on this consideration, the options that were presented in the analysis should be screened one by one to find out which of them are ruled out by current infrastructural limitations that are not likely to be overcome in the near future. Table 3.24 suggests an appropriate series of points to be checked through at this stage.

Financial Resources

Another factor that is likely to limit the choice of possible actions items are the financial resources available to the service operation. Therefore, it is essential to identify the resources that are currently available and those that are predicted to be available in the near future. Also, it is useful to think about which resources could be shifted internally, if needed, and what would be the impact. Based on this financial consideration, the options that were presented in the analysis should be screened one by one to find out which of them are ruled out by lack of current and/or predicted financial resources (that probably cannot be compensated otherwise, for instance by shifting financial resources internally). Table 3.25 presents a brief but significant checklist for this sub-step.

Table 3.24   Checklist – infrastructure

• What is the current infrastructure able to deliver the services to customers and what are the infrastructural limitations?
• Could any of these infrastructural limitations be overcome and how?
• Going through all service standards one by one, which of the options presented in the analysis are ruled out by current infrastructural limitations that are not likely to be overcome in the future?

Table 3.25   Checklist – financial resources

- What currently are the financial resources available to the service operation and what are those predicted to be in the future?
- Going through the service standards one by one, which of the options presented in the analysis are ruled out by lack of current and/or predicted financial resources?

Human Resources

A tremendously decisive factor in the delivery of services are the human resources available to the service operation. If customer contact staff is unfriendly, not customer-focused, not empowered, not well trained, and/or not motivated, the service operation will be out of business before long if customers have got the choice to do without it (which may not be the case in the public sector or within larger companies).

If the personnel delivering the services to customers, on the other hand, are always friendly, show honest respect and appreciation for their customers, are well trained, highly skilled and competent, empowered and highly motivated, the service operation is likely to be doing very well.

As discussed in Chapter 2, Zeithaml and Bitner (2002) suggest the following human resources strategies to ensure the right personnel are available to the service operation:

- *Hire the right people* – by competing for the best people, hiring for service competencies and service inclination, and being the preferred employer.
- *Develop people to deliver service quality* – by training for technical and interactive skills, empowering employees, and promoting teamwork.
- *Provide the needed support systems* – by developing service-oriented internal processes, providing supportive technology and equipment, and measuring internal service quality.
- *Retain the best people* – by including employees in the company's vision, treating employees as customers, and measuring and rewarding strong service performers.

The question needs to be asked: 'Who are the human resources currently available to the service operation and what are their qualifications, experience, potential and motivation?'. Also, it is important to know whether it is possible to actually hire new people; in other words, who has the authority and/or financial resources to do so, and how long does the process of hiring somebody take in the service provider?

Based on these human resources considerations, the options that were presented in the analysis should be screened one by one to find out which of them are ruled out by lack of suitable personnel that cannot be overcome by hiring new people and/or training the existing resources. Table 3.26 summarizes these points as a checklist.

Table 3.26   Checklist – human resources

- What are the current human resources of the service operation?
- What are the qualifications, potential, experience and motivation of those people?
- Have you got the resources and/or authority to recruit additional team members?
- Going through the service standards one by one, which of the options presented in the analysis are ruled out by lack of suitable personnel?
- Could this lack be compensated for by hiring new people and/or training the existing team members?

Action Items

In order to come up with a concrete set of action items, the service standards should now be examined one by one and it should also be checked to ensure that none of the options produced in the previous step have been ruled out by any of the above considerations. This activity can be visually enhanced if all options are written in a spreadsheet or matrix against the considerations described above. Then the question must be asked as to which of the remaining options or combination of options seem to be the best choice under the circumstances and why.

In case all options for an individual service standard have been ruled out, it is possible that there might be a compromise solution that limits the damage to both customers and the service operation.

Finally, a list of best-choice concrete action items per service standard must be generated, based on the above considerations, and an explicit statement made that identifies the desired effect of each action item.

Table 3.27 provides an appropriate checklist.

Table 3.27   Checklist – action items

- Going through the service standards one by one, which of the options presented in the analysis have not been ruled out by one of the previous considerations?
- Which of those or combination of those seem to be the best choice under the circumstances?
- For service standards where all options have been ruled out, is there a compromise solution that limits the damage to the customer and the service operation?
- Generate a list of best-choice concrete action items based on the previous considerations.
- State the desired effects and risks of each of those action items.

In Line with the Expanded Marketing Mix?

Airline operations
A traditional airline needs to cut costs in order to return to profits and at the same time seeks to increase flexibility to cope with fluctuations in demand. One option to achieve this could be to reduce cabin staff on flights, thereby freeing personnel for additional flights (increasing flexibility) and reducing personnel costs per flight. When evaluating all options identified, however, it needs to be carefully considered whether or not this option is in line with the airline's overall marketing mix. In particular, the promotion, price and people components may not be in line with such an option. For instance, if the airline justifies higher prices than low-cost providers because the on-board service is excellent and carried out by highly qualified and trained flight attendants, then a reduction in cabin staff is likely to undermine that position.

Airport operations
A small airport offers services to two low-cost airlines at very low fees, which was one of the most important reasons for both companies to operate from this airport. In order to be able to charge those low fees, the airport has to improve its cash flow somewhere else. One of the options considered is to delay the investment in additional ground support equipment. This option may be in contrast with the promotion component of the airport's marketing mix, since customer airlines were promised turnaround times of under 30 minutes, which currently is only possible with one aircraft at a time.

Aircraft manufacturing operations
An internal service provider delivering cutting-edge development methods and tools needs to perform better on the support quality delivered to customers and, at the same time, cut costs. One of the options evaluated is to increase the number of sub-contracted resources bought in for service delivery. Although this option seems to result in lower personnel costs and avoids long-term contractual obligations, it may not be in line with the people component of the marketing mix, stating that experienced and highly qualified long-term employees will be delivering the services offered.

In this step, having derived concrete action items from the analysis done during Step 4 and some additional considerations, the next step consists of closing the cycle by implementing and following up the derived action items.

Step 6 – Closing the Cycle

The sixth step (see Figure 3.13) is about closing the service quality cycle by implementing and following up the derived action items.

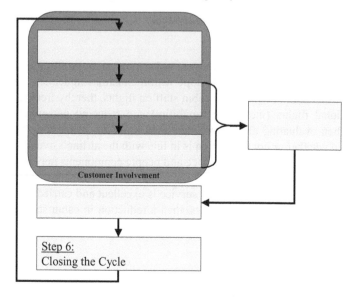

Figure 3.13   Closing the cycle

In order to close the service quality cycle a number of sub-steps are proposed here (see Figure 3.14). The starting point of Step 6 is the set of action items that was derived in the previous step. To implement and follow up the action items, a number of issues should be considered that are described in the sub-steps *timing, dealing with resistance, priority* and *communications to customers*.

Looking at the whole cycle, it also has to be decided how often (that is, in what time intervals) the cycle should be repeated and whether the customer involvement phases of Steps 1, 2 and 3 could not be combined in some form to minimize disturbing customers, once the cycle has been established for the first time.

Action Items

Looking at the set of derived action items (including a statement per action item as to how each action item is expected to improve the situation) as the starting point of this step, certain aspects have not been considered yet, such as timing, priority, resistance, communications to customers, and who is actually carrying out each action item. Nor has it been decided how each action item will be followed up to ensure that the expected results actually materialize. Those issues are addressed below.

Dealing with Resistance

No matter what changes are implemented, there is always the potential for resistance against them. Reasons for and strategies to overcome resistance were discussed in

Step 6:
Closing the Cycle

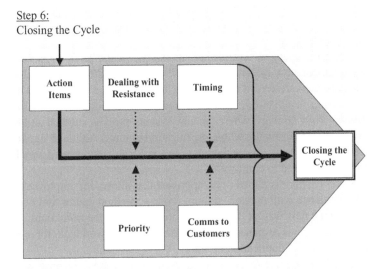

Figure 3.14 Step 6 – Closing the cycle

Chapter 2 in more detail (see Table 2.5). The set of derived action items is likely to cause some sort of internal and external resistance, be it by a small number of isolated individuals or entire teams.

Whatever the causes for resistance may be in a specific case, resistance and the efforts to overcome or resolve resistance almost always represent additional work and a deterioration of the working climate and trust, at least temporarily, all of which is not in the interest of the service operation. A framework of how major change can be managed effectively was presented in Chapter 2 (Table 2.6).

What is in the interest of the service operation, however, is to pre-empt resistance by communicating to, involving and supporting affected employees and customers at an early stage before resistance arises. There are two principal outcomes of such a policy. Either the affected employees or customers know about the changes early, feel involved and supported and, hence, are not opposed to the changes and even promote them if they are convinced; or there are really weak points in the changes ahead, in which case the decision can still be modified or taken back before much damage is done.

The main disadvantage of such an approach is that it is very time-consuming to try and convince potential resisters and allow for their participation. Also, in some business cultures, it may be seen as weakness in a manager if subordinates are asked their opinion. On the positive side, participating potential resisters can contribute to improve the decision and/or, once they are convinced of the usefulness of the changes ahead, they will support them.

Influencers and/or opinion leaders should be identified and targeted as early as possible. Once they are convinced they are likely to help promote the changes and function as change agents.

In case of major changes, several alternative *implementation strategies* are widely used to implement new systems, processes, methods and/or tools in companies. Although these were described in Chapter 2, they need to be re-examined here:

- The *big bang* strategy, where the new system, process, method and/or tool is implemented across the whole organization at once. Although duplication of work is saved using this method, it is very risky and represents a very high and sudden workload to employees.
- *Parallel running* is a lower-risk approach that allows for employees building up confidence with the new system, process, method and/or tool because the old and the new way of doing things are running at the same time. However, this approach is expensive and means a higher workload for employees because there is duplication of work.
- The *phased approach* represents a useful compromise insofar as it allows for modifications in the light of first experiences and does not expose the whole company at once if there are still major problems with the new system, process, method and/or tool. Within parts of the company either of the above approaches can be used, until – step by step – the implementation is completed.
- The *pilot study* approach seems suitable for high-risk projects, but is time-consuming and expensive.

Finally, customer contact employees must be sufficiently empowered to take immediate actions if they receive complaints from customers. Time is absolutely crucial if customers have to be turned around from being utterly disappointed to accepting a good recovery from the service operation, if things go wrong. Letting – within reason – the contact employee manage the situation on the spot can save

Table 3.28   Checklist – dealing with resistance

- Consider what internal resistance could be expected.
- Keep your team sufficiently involved and informed along the entire service quality cycle.
- Take time to convince them, openly listen to their ideas, worries, doubts or fears, and seriously take them on board.
- Consider what external resistance could be expected.
- Identify which customers are influencers and/or opinion leaders and/or could be used as change agents.
- In the case of major changes, choose a suitable implementation strategy.
- Make sure your team members are sufficiently empowered.

Table 3.29   Checklist – priority

• Decide on the priority of individual action items. • If there are not sufficient resources to implement all action items at once, decide which should be addressed first, based on their priority.

a lot of time. Also, the customer will immediately feel taken seriously. In order to improve the effectiveness and the efficiency of complaint management and recovery, they must be conducted systematically and recovery must be pre-planned so that customer contact employees are able to react swiftly.

In the case of major, far-reaching changes across larger companies it is highly recommended to consider the 'eight-stage process of creating major change' as proposed by Kotter (1996) (see Table 2.6). For most changes in connection with the step 'closing the cycle', however, the pragmatic approach discussed above will be sufficient.

Table 3.28 summarizes the significant points regarding this sub-step.

Priority

All action items derived in the previous step of the service quality cycle should now be put in an order of priority. It is important to establish this order of priority from a customer's perspective – without actually asking them – unless it is impossible for that choice to be made internally. This could be the case if there were two or several mutually exclusive action items.

If there are not sufficient resources available to tackle all action items at once, which will often be the case, it should be decided – based on the priority list – which of them will be implemented first. Table 3.29 provides a short but important checklist for this stage.

Timing

Depending on the resources that are needed to carry out each individual action item, the availability of those needed resources must be checked. Also, particularly when the active participation of customers is required to implement an action item, the customers' schedule must be considered. Customers' relevant deadlines and/or milestones need to be taken into account to best serve their timing needs. Based on these considerations and the priorities discussed in the previous sub-step, it should be planned when each individual action item can or has to be implemented, as summarized in Table 3.30.

Communications to Customers

From a service operation's viewpoint, all communications to customers are crucial insofar as they contribute to customers (mostly) subconsciously building up both

Table 3.30   Checklist – timing

- Consider when the resources that are needed to implement each action item will be available.
- Consider the customers' schedules, especially if their direct involvement is needed to implement a specific action item.
- Consider when there are relevant customer deadlines and/or milestones.
- Based on the above considerations and the priority allocated to each action item, schedule when each one of them will be implemented.

---

their expectations and perceptions of the services provided. If communications are consciously managed by the service operation, it is possible to significantly influence both customer expectations and perceptions of the services delivered. As was shown in Chapter 2, a great source of potential customer dissatisfaction lies in the gap between what customers expect from a service (based on different factors that influence these expectations) and how they actually perceive the service delivered.

Therefore, it is recommended here – by means of communication – to proactively influence both customer expectations and perceptions, even exceed customer expectations where possible and deliver a consistent image between the selected service standards and the communications of the service operation.

In order to *proactively influence customer expectations*, realistic and accurate promises should be made that reflect the service actually delivered (explicit service promises). Also, it should be assured that service tangibles accurately reflect the service provided (implicit service promises).

Customers often subconsciously look for tangible side products of services to compensate for the property of intangibility of services. Such tangibles may be business cards, reports, bills, e-mails, etc., from the service operation's employees, etc. They give the customer the feeling they can actually 'touch and feel' the service, although the service itself is intangible. Hence, those tangibles must be taken seriously to leave a good impression.

Next, customers ought to be 'educated' to understand their roles and perform them better. For instance, if customers have to perform one or two steps themselves in order to receive authorization to use a new tool after they were trained, then they must be educated in their role to request authorization themselves. Otherwise, there is a risk that they will blame the tool or the trainer for their still not having access to it.

Very importantly, influencers and opinion leaders in the customer organization have to be identified and marketing efforts concentrated on them because they tend to act either as sources for resistance if they are not convinced of the usefulness of the services, or, on the other hand, as change agents who effectively promote the services offered inside the customer organization if they are convinced, at least by spreading positive word-of-mouth information. Since resources to talk to customers will be limited, it is also more efficient to aim communications at these key players.

Last, customers ought to be told when service provision is higher than what can normally be expected so that predictions of future service encounters will not be inflated.

In order to *proactively influence customer perceptions*, customer satisfaction should be aimed for in every service encounter. Each and every meeting, even short crossings in the building or outside, will contribute to build or maintain customer perceptions about the service operation. Therefore, each encounter must be seen as an opportunity to improve customer perceptions of the service operation. Also, evidence of service should be managed proactively to reinforce perceptions. Positive achievements should be communicated correspondingly. Finally, by using (positive) customer experiences, (positive) image will be reinforced over time.

Customer expectations can be exceeded by under-promising and over-delivering, as well as positioning unusual service as unique, not the standard. The former approach takes place before a service is delivered, the latter after a service was delivered at a better level than expected. For example, if a supplier promises to deliver a specific service by a certain date, but under-promises (perhaps that date is very realistic and the service is actually supplied one day earlier), then the customer will see his/her expectations as exceeded because there was over-delivery. If the supplier then explains to the customer that in this case it could deliver earlier than promised but that this was an unusual situation, then the supplier pre-empts inflation of customer expectations for future service deliveries.

To conclude, it has to be emphasized that all communications to the customer have to correspond to the service standards selected and that the service operation has to deliver to those standards consistently over time.

It takes a long time to achieve a good standing and reputation amongst customers, but only a little time to destroy it by sending out the wrong communications. Table 3.31 provides a checklist for this and the other points of this essential aspect.

Table 3.31   Checklist – communications to customers

- Proactively influence customer *expectations* by making realistic service promises (both explicit and implicit), educating customers about their role, identifying and targeting influencers and opinion leaders, as well as telling customers when service provision is higher than what could normally be expected.
- Proactively influence customer *perceptions* by aiming for customer satisfaction in every service encounter, actively managing service evidence, and using customer experiences to reinforce images.
- Exceed customer expectations by under-promising and over-delivering, as well as positioning unusual service as unique, not the standard.
- All communications must be in line with the service standards selected.

Closing the Cycle

It is very important to close the cycle for a number of reasons. For instance, failure to do so may mean that a supplier deduces it has good standing in the eyes of its customers and possibly decided on some useful action items. Customers, however, may be disappointed if their participation in any measurements does not result in improvements of their situation.

Also, from an economical viewpoint, it costs some time and money to implement the suggested method for the first time, but hardly anything to keep the cycle going again and again. There will always be slight adaptations, but the efforts needed to start service quality management from scratch every time you need a basis for decision-making would be far greater.

Implementing the method will show to be very effective in the short run, whereas *closing the cycle* brings about efficiency and enables you to proactively control service quality over time. This process is summarized in Table 3.32, but it is important to look at the detail first.

In order to close the cycle – to actually implement and follow up the derived action items – a series of issues should be addressed. First, based on the above considerations, one or several persons should be assigned to implement each specific action item. Ideally this would be the employee that is best suited to perform the task. In light of priority and availability issues, however, a different choice may have to be made. Second, any steps needed to pre-empt internal and/or external resistance to the changes brought about by the action items decided should be clearly stated. Third, it should be described for each action item what is the planned timeframe for its implementation, its relative priority and how it will be communicated internally to employees and externally to customers. Fourth, the decision of how each action item will be followed up over time should be taken, to ensure that the expected improvements really materialize. Usually, this follow-up should take place in the

Table 3.32   Checklist – closing the cycle

- Based on the above considerations, allocate each action item to one or several persons who are responsible for carrying it out.
- Clearly state any steps needed to pre-empt internal or external resistance.
- Going through the list of action items decided, clearly describe the timeframe for each action, its priority, and how it will be communicated internally and externally.
- For each action item, decide how it will be followed up over time to ensure that the expected improvements really materialize.
- Decide in what time intervals the service quality cycle should be repeated and whether the customer involvement parts of Steps 1, 2 and 3 could be combined in some form, in order to minimize disturbance for customers, once the cycle has been established.

Internal and External Resistance?

Airline operations
Some of the action items decided may involve shifts of flight crew and cabin staff deployment, regarding working hours and locations, so that the operation achieves higher levels of flexibility, but specific groups of employees and possibly unions may not agree with some of the steps taken. In such cases, a reasonable solution must be found, of course, that is acceptable to all because the commitment of everybody concerned is essential. One good approach seems to be early involvement of representatives of the groups of employees that are immediately concerned. External resistance may be an issue but often it appears to be of lesser importance than internal resistance against the measures taken because the very goal of these is to improve the services delivered to passengers.

Airport operations
Airlines are less likely to resist against improvement steps taken by airport operations, as long as safety, fast turnaround of their aircraft and their cost basis are not compromised. Passengers are not likely to resist if those steps result in faster throughput at check-in and security checkpoints and service levels are improved. Shops and restaurants in the airport as well as suppliers' own personnel and unions may resist longer working hours. Neighbouring communities are likely to resist additional flights during the night and there may even be legal battles ahead against such moves.

Aircraft manufacturing operations
Serving external customers often requires high flexibility of the teams concerned and steps to further improve that flexibility may not be welcomed by front-line employees and the relevant labour union. As mentioned above, early involvement of the relevant employees is essential to come up with measures that are supported by them, which is crucial because it is they who have to face the customers. Serving internal customers (for instance, by providing a design department with a new and more powerful IT infrastructure), internal resistance within the service provider will be less of an issue. Rather, their internal customers need to be convinced of the usefulness of the change.

form of the service quality cycle being repeated, be it as a whole or in part (see discussion below). Finally, it ought to be decided in what time intervals the service quality cycle should be repeated and whether the customer involvement phases from Steps 1–3 could be combined in some form, in order to minimize disturbance for customers once the cycle has been established. For instance, if the service standards and their relative importance seem to be relatively stable over time, the customer involvement phases of Steps 1 and 2 could be left out during one or several cycles. However, it should occasionally be verified with individual customers (informally) whether this assumption still seems to hold true.

Having gone through the last step of the service quality cycle method (that is, how the cycle should be closed to allow for control of service quality over time), the next section summarizes all major benefits inherent in using the suggested method.

Summary of Benefits

The following table (Table 3.33) contains a summary of the benefits inherent in applying the method suggested in this chapter. Those benefits are grouped into the individual steps of the service quality cycle they result from (see Figure 3.1) and the overall method. Only major benefits are stated here that should give the reader an overview of why it is worthwhile using the method.

After presenting a step-by-step method to implement an effective yet efficient system to control service quality over time in this chapter, the following chapter describes and critically evaluates a recent case study from the aerospace industry that offers useful insights in an example implementation of the method. The appendix at the end of this book offers a ready-for-use summary checklist containing all individual steps of the method.

Table 3.33   Summary of benefits

Source	Benefits of using the proposed method
From Step 1	– A set of validated, relevant, achievable and valued service standards is created that can be used to set the service operation's targets internally and to communicate those to customers, thereby influencing their expectations from the service operation.
	– All employees and key customers involved in the generation of these standards are most likely to feel 'ownership' of them and support them because they contributed to generating them.
From Step 2	– The relative importance that customers attach to each of the service standards is known.
	– If there are significant differences in the perception of this relative importance as expressed by individual customer teams or segments, this can be assessed visually from the spreadsheet.
From Step 3	– It measures what really matters to the customer and what the service operation really is trying to achieve. In other words, the criteria for this measurement are very meaningful and relevant, both to customers and the service operation.
	– By using a suitable mix of means to collect feedback, the likelihood of receiving carefully balanced and highly relevant feedback (both quantitative and qualitative) is increased.
	– A positive side-effect is that by asking customers to rate the service operation's performance against the service standards, those very standards are regularly communicated to the customers, which then influences their expectations from the services delivered.

Source	Benefits of using the proposed method
From Step 4	– The importance-performance matrix helps to visually present customer perceptions of how well the service operation is performing against each service standard and where there is room for improvement (that is, it gives a snapshot of the current situation).
	– The overall situational context is taken into account, in parallel with the internal and external factors that can be influenced by the service operation and customers, in order to avoid jumping to premature conclusions about the current situation.
	– A balanced analysis per individual service standard is produced that also offers alternative options, including their advantages and drawbacks, thereby greatly enhancing decision-making in the following steps.
From Step 5	– The options produced during the previous step are systematically screened for feasibility and suitability.
	– The outcome of these considerations are a set of well balanced best-choice action items that fit in the situational and strategic context, and take into account a variety of constraints of the service operation.
	– For each action item that is derived, the desired effects are explicitly stated, which allows for following up and finding out whether the decisions actually led to the expected results.
From Step 6	– For each action item, timing, priority and availability of resources issues are discussed and taken into account, when deciding which action item is implemented when and by whom.
	– Internal and external sources of potential resistance against the changes linked to the implementation of the set of action items are identified and resistance is systematically pre-empted.
	– Clear and consistent communications to customers are enhanced that proactively influence both customer expectations from and perception of the services delivered, leading to higher levels of overall customer satisfaction.
Overall method	– The suggested method, the service quality cycle, enables service providers to effectively and efficiently improve and control their customer-perceived service quality over time.

Chapter 4

Case Study: Aircraft Manufacturing

Chapter Summary

The previous chapter has described a step-by-step method of how service quality can be improved and controlled over time, the 'service quality cycle'. The purpose of this chapter is to describe a practical example of the implementation of the proposed method. This will be done in the form of a recent case study that took place in the European aerospace industry, more precisely in a leading aircraft manufacturing company, from January to December 2004.

In order to do so, the context of the case study will be given first, including a brief description of the customers, the service provider, and the organization in which it is embedded. Second, the status before the implementation of the service quality cycle is summarized. Third, the actual implementation and repetition of the cycle over time is presented. Fourth, the status at the end of the 12-month period is described, and, finally, the lessons learned from this case study are summarized.

Context of the Case Study

This case study describes the implementation of the method (as proposed in this guide) in a leading aircraft manufacturing company. Within that company, the supplier organization delivering process, method and tool support services for systems engineering is a multinational group that is organized and run transnationally in France, the UK, Spain and Germany. In this case study, it is referred to as the 'Support Services Group' (see Figure 4.1). For confidentiality reasons and to simplify, the real names of the different units and sub-units are not used in this book.

The main objective of this group is to enable, enhance and facilitate the work of design and development teams by providing them with best practice, state-of-the-art *processes, methods* and *tools* (supporting those processes and methods) so that they can best concentrate on their respective areas of expertise. The services that are provided are aimed to reduce the time needed for development, improve the quality of the products developed, and reduce the overall development costs the company incurs.

The Support Services Group is composed of an overall 'Management', a number of different 'Capability' teams and a number of 'Program' teams. Management and all teams are transnational, the latter two being organized in a matrix structure.

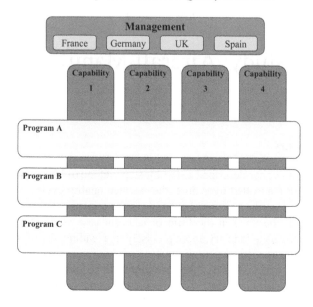

Figure 4.1   Organization of the Support Services Group

Management consists of the manager of the Support Services Group as well as one site representative of the group for each country (France, Germany, UK, Spain). Each *Capability team* consists of one leader and many process, method and/or tool experts in one specific field of expertise such as requirements engineering, interfaces management or configuration management. The members of each Capability team are located across all major sites of the company. It is the responsibility of each Capability team to foster the respective capability, keep it up-to-date, and permanently look for best practice in their field. Also, some members of each Capability team are used to actually deliver the relevant support services to the customers within one or several Program teams. *Program support teams* are delivering the support services to all design and development teams of new aircraft programmes (their customers) across all relevant sites of the company. Each Program team is basically arranged as one service operation per country, so that support is delivered in most cases nationally but in a harmonized way across all sites concerned. The resources used to deliver the services are mostly taken from the relevant Capability teams.

Within that transnational Support Services Group, more specifically within one Program team, one transnational team of process, method and/or tool specialists is managed in Germany and is considered as the 'service provider' in the present case study (in other words, they form the service operation). The service provider (see Figure 4.2) enhances and facilitates the work of their customers by delivering best-practice process, method and tool support (training, coaching, helpdesk support) in such areas as requirements engineering, interfaces management, data modelling,

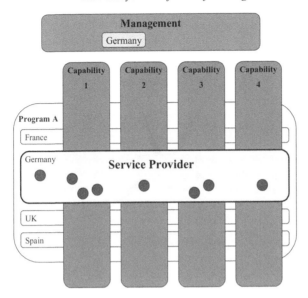

Figure 4.2 **Organization of the service provider**

software simulating, validation and verification activities, as well as configuration management.

Correspondingly, the human resources used to form the service provider are mostly taken from the locally based (Germany) members of the Capability teams. The service operation is managed by a so-called Capability Integrator (general focal point of the service provider) whose task it is to analyse demand, plan capacities to meet that demand, as well as to coordinate the support services. Reporting takes place on a weekly and a monthly basis to the management of the Support Services Group and that of customer teams simultaneously.

The financial resources available to the service provider are allocated after yearly budget negotiations between customer teams and the Support Services Group. It was decided to change this insofar as from the year after the case study those negotiations were going to take place directly between the local service provider and those customer teams located in the respective country. Hence, the service provider can be seen as an internal service operation, organized as a cost centre.

The present case study covers a period of time of 12 months and all support services that were delivered to the customer teams by the service provider during that period.

The 'customers' of the service provider are design and development teams (and their individual members) for systems and structure components within a current aircraft development programme of the company. The teams in question are composed mainly of systems engineers and designers with different cultural, academic and professional backgrounds who are located in many major and minor sites of the

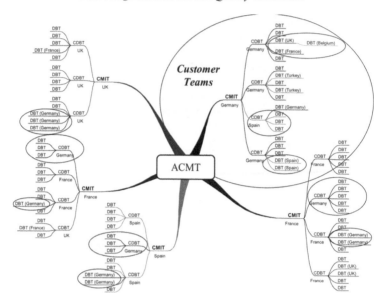

Figure 4.3   Organization and locations of customer teams

company in Germany, France and the UK. Some of their teams (belonging to partner companies) were also located in Turkey, Spain and Belgium, but not directly or fully supported by the service provider.

Within the customer organization (that is, the new aircraft development programme), customers are organized in ACMTs (Aircraft Component Management Teams), which in turn are sub-divided into CMITs (Component Management Integration Teams) that are composed of CDBTs (Component Design Build Teams) and their subordinated DBTs (Design Build Teams). Depending on where specific know-how is located, those teams and sub-teams are situated across all different sites and even at sites of sub-contractors of the company.

Confusingly, as it may seem at the first glance, customer teams at ACMT, CMIT, CDBT and DBT level are rarely located at the same site but rather are likely to be spread across different sites. Figure 4.3 gives a simplified overview of a typical organization with locations of customer teams and sub-teams, as it was found in the case study.

Having looked at the context of this case study, including organizational aspects of both the Support Services Group and customer teams, the next section will describe the status at the beginning of the case study with respect to service quality management.

Status Before

At the beginning of the present case study, only one large customer team (located at a major site of the company in Germany with many sub-teams spread across Europe) was supported in some of the major development methods and tools on site. All other teams in Germany were not supported locally but from France, resulting in individual teams not being supported at all or at a very low level only. Many customer team members were relatively new to their job for the ongoing aircraft development programme in question. Time pressure on customer teams was quite high, with engineering deliverables such as requirements and design specifications having to be produced in time and in the right quality at specific programme milestones. Also, those teams that were supported were only supported in one of the capabilities at hand, mainly because at the given stage of the programme, the capability in question was the most critical one. Resources that were allocated from the relevant capability team to the program team in Germany were very limited.

With respect to proactive service quality management, customer-driven service quality standards were neither defined nor communicated, nor was customer-perceived performance against those standards measured in any way, thus preventing systematic improvement and control of service quality.

Although the level of commitment and motivation of the few resources available to deliver the limited range of support services was very high, the reputation of the Support Services Group amongst customers in Germany was relatively low. This was, at least in part, attributable to the fact that high expectations had been raised by management promising excellent and far-reaching support services to customer management, but not actually allocating the resources nor the authority necessary for the program team to meet these expectations. Also, there was a lack of clear strategic guidelines and directives from the Support Services Group, and it had only just been decided to organize the program team under consideration per country (as presented in Figure 4.2).

Figure 4.4 gives a simplified overview of the coverage of specific capabilities (that is, the range of support services offered) over some of the potential customers who are addressed and served in Germany at the beginning of this case study. As can be seen in Figure 4.4, only a relatively low percentage of customers were covered and only a small part of the range of support services offered to customers at the beginning of the case study.

Having looked at the status with respect to service quality management at the beginning of the case study, the following section will describe in detail the implementation of the service quality cycle (the method suggested in Chapter 3) during the course of the case study.

Implementation of the Service Quality Cycle

This section describes the implementation of the method suggested in Chapter 3 of this book, as it was conducted by the service provider under consideration from

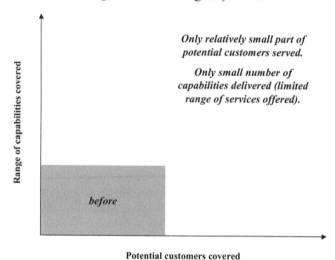

Figure 4.4   Support services coverage in Germany (status before)

January to December 2004. As the method explicitly allows for, not all steps have been carried out every time or with the same emphasis.

The implementation of the service quality cycle as carried out in the course of the present case study, and as described here, covers the *following steps*:

- *First cycle run:*
    - definition of internal service standards
    - consolidation and ranking of external service standards
    - preparation of a performance measurements questionnaire
    - performance measurements, March 2004
    - analysis of the current situation, March 2004
    - derivation of action items, March 2004
    - closing the cycle
- *Second cycle run:*
    - performance measurements, May 2004
    - analysis of the current situation, May 2004
    - derivation of action items, May 2004
- *Third cycle run:*
    - performance measurements, July 2004
    - analysis of the current situation, July 2004
    - derivation of action items, July 2004
    - decisions on next steps in using the method

- *Fourth cycle run:*
  - performance measurements, December 2004
  - analysis of the current situation, December 2004
  - derivation of action items, December 2004
  - decisions on next steps in using the method.

First Cycle Run

As described in Step 1 of the service quality cycle, and based on the five performance objectives, the first step put into practice was the internal definition of service standards that could later on be presented to customers and serve as the basis for customer-defined or at least customer-driven (external) service standards. In doing so, a set of internal service standards (that related to quality, speed, dependability and flexibility) were formulated and further discussed and refined with experienced customer contact employees of the Support Services Group in January 2004 (see Table 4.1).

The cost performance objective was not followed up because the service provider could not directly influence it, since it was part of the same company and fixed hourly rates were applied for the services delivered.

Each of these internally defined service standards was intended to be as self-explanatory as possible, but they all were also defined in detail to avoid misinterpretations by customers and own staff alike.

Following the establishment of internally defined service standards, customer involvement both to consolidate service standards and have them ranked (Steps 1 and 2) was prepared. A combined approach (that is, customer involvement parts

Table 4.1   Initial service standards (internally defined)

Internally defined service standards	Related performance objectives
Support capacity must match customer demand	Dependability Flexibility
The quality of training, coaching and helpdesk support must be high	Quality
Presence of support team members in customer teams is high	Speed Dependability
Presence of support focal point in relevant customer meetings is high	Speed Dependability
Reactions to support relevant problems are immediate	Speed
Support delivery is speedy and flexible	Speed Flexibility
Support services are highly flexible	Flexibility
Customer suggestions are followed up seriously	Dependability

Table 4.2   Consolidated and ranked service standards (externally defined)

Customer-driven service standards	Related performance objectives	Relative importance
(1) The *speed of reaction to problems* is very high	Speed	1
(2) The *training quality* is very high	Quality	4
(3) The *coaching quality* is very high	Quality	3
(4) The *helpdesk support quality* is very high	Quality	2
(5) *Promises* made are kept	Dependability	5
(6) The *overall impression* by support team members delivering services to the customer is very positive	Quality	6
(7) Support team members flexibly adjust to *working hours* and *locations* of the customers	Flexibility (Delivery)	7
(8) The service provider flexibly offers *new services* as the customer need arises	Flexibility (Product/Mix)	7

for both steps) is only recommendable if changes to the discussed standards are unlikely, otherwise the rankings may turn out to be obsolete and have to be repeated, wasting time and, even worse, annoying customers.

In the given case, first discussions with individual customers led to refined, self-explanatory service standards until a certain status of maturity was reached and the current set of service standards was considered stable enough to ask remaining customers both to consolidate and to rank them at the same time (combined approach).

In order to find out about the service standards' relative importance as perceived by customers, a scale of importance needed to be chosen. The nine-point scale of importance as described in Chapter 2 (see Table 2.3) was selected. Both, the set of externally consolidated service standards and their relative importance (rounded values) are presented in Table 4.2.

The form of customer involvement used to define and rank service standards were interviews with individual customers and focus groups with several customers at the same time, usually coming from the same customer team.

In order to prepare performance measurements by customers, an initial questionnaire was prepared (see Table 4.3) that should also serve as the basis for interviews to measure the supplier's own performance, as perceived by the customers, along those service standards defined previously.

The scale of performance chosen for these measurements was a scale from 1 ('I do not agree at all') to 10 ('I fully agree'). The questionnaire was sent out to all

Table 4.3   Support services performance measurements questionnaire

Dear Customer,

You can help us achieve more customer orientation and better adjust our service offerings to **your** needs by taking a minute and filling out the table below.

Please only rate statements that you feel confident with and feel free to send us back this form either by e-mail or anonymously by fax or internal post.

Thank you very much for your participation!

To what degree do you agree with the following statements?	Performance 10 – I fully agree. 1 – I do not agree at all.									
Service standards	10	9	8	7	6	5	4	3	2	1
(1) The *speed of reaction to problems* is very high.										
(2) The *training quality* is very high.										
(3) The *coaching quality* is very high.										
(4) The *helpdesk support quality* is very high.										
(5) *Promises* made are kept.										
(6) The *overall impression* by support team members delivering services to the customer is very positive.										
(7) Support team members flexibly adjust to *working hours* and *locations* of the customers.										
(8) The service provider flexibly offers *new services* as the customer need arises.										

Do you have any additional comments, improvement suggestions or ideas we should take into account to improve and maintain high service quality?

customers identified as an attachment to an e-mail. Customers were given the choice to send back the questionnaire by e-mail (easiest, but not anonymous), fax or letter (more hassle and slower, but anonymous).

The overall reply rate to the questionnaire was under 20 per cent and all replies that were received were sent back by e-mail within a period of up to two weeks.

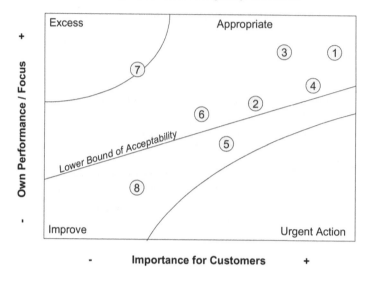

Figure 4.5   Performance against service standards, March 2004

Individual interviews were a good compensation for the lack of participation amongst customers. Results from both questionnaires and interviews did not show significant differences between individual customer teams so that the average values were used to visualize performance.

For future performance measurements, it was decided to use a very similar questionnaire but as part of one e-mail, so that customers did not have to double-click on the attachment to open the questionnaire (which many did not even bother to do), but they could directly see it when going through their e-mails. Also, the questionnaire would be sent in both languages, English and the local language (German), next time, with little flags positioned in the e-mail to directly indicate that fact to customers.

The position of each individual service standard in the importance-performance matrix after the first performance measurement in March 2004 is presented in Figure 4.5, using the average importance values allocated to each service standard and the average performance values given by customers. Each service standard can be found in the figure, indicating via its position in the matrix what was its relative importance allocated by customers (*x* axis) and the currently perceived performance of the service provider against the standard in question (*y* axis).

Table 4.4 covers the discussion and analysis of each of the service standards as well as the current situational context, in order to form a sound basis for operational and strategic decision-making in the form of concrete action items later on.

Both the outcome of the analysis (including the discussion of options) and the derivation of action items are likely to be different in different situations, and/or if

Table 4.4   Analysis of the current situation, March 2004

Subject	Analysis
Situational context	By the time of the first performance measurement in the given case study, not all customers currently served in Germany had been covered and the service provider was just in the ramp-up phase, with only about half the current staff being part of it.
	Also, not all of the customers served at that time, mostly those customers from one of the major sites of the company (that had not previously been covered by the service provider), had been in direct contact with any member of staff, but usually only the general focal point of the service provider (the Capability Integrator).
	Some training elements were still outsourced and performed by external resources. Still, customers perceived those training elements to be the responsibility of the service provider because they covered tools that were part of the usual scope of services offered.
	In the past, some current customer teams were supposed to be supported by a similar service provider (within the Support Services Group) from France because those teams' management was located there. This resulted in poor or no support at all for a variety of reasons, mainly language and cultural barriers.
	In the months before the case study, there were some cases of over-promising and of behaviour (by individual members of the Support Services Group that are not part of the service provider) that was perceived by some customers to be arrogant. This resulted in an initially overall negative attitude towards the service provider amongst many customers. The belief that the service provider could actually help the development teams to save time and increase product quality was not widespread at the beginning of the case study. Only the fact that it seemed to be new members of staff who started up the service provider's activities for the new aircraft development programme in question helped to overcome this initial ill-feeling amongst those customers with negative attitudes towards the Support Services Group.
	The budget situation for the current year was not very satisfactory, because the financial resources available to the service provider had been determined under tremendous pressure to cut overall budgets. Also, they did not take into account new additional processes, methods and tools to be covered from this year on, nor the increased number of customers due to the new support organization per country.
	Budget negotiations between customer teams and the Support Services Group took place on a yearly basis and it was decided that from the coming year onwards those negotiations should be conducted directly between the local service provider and those customer teams located in the respective country.
	There were no clear contractual obligations in place such as service agreement letters or services contracts between customer teams and the service provider.

Subject	Analysis
(1) The *speed of reaction to problems* is very high.	The current position in the importance-performance matrix is in the appropriate zone and indicates that this service standard is perceived to be the most important one of the entire set of service standards. Also, customers perceive the service provider to perform very well against this standard. Reasons for this position are likely to be that customer contact staff of the service provider are highly motivated and always leave the impression that they will immediately try to resolve any customer problems. This position can be considered to be perfectly all right and there does not seem to be the need for change. *Option:* The efforts within the service provider that have led to this good position need to be maintained and employees positively reinforced, in order to pre-empt complacency.
(2) The *training quality* is very high.	Training quality lies in the appropriate zone. Its current position indicates that customers perceived this service standard to be of medium to high importance but less important, for instance, than coaching or helpdesk support quality. The service provider was perceived to perform adequately against this service standard, but its position lies just above the lower bound of acceptability. The reason for this result seems to be that the majority of the training offered at that moment in time was carried out by a subcontractor who was not familiar with specific customer processes and methods, but merely knew well one specific tool that they supplied to the company. In other words, although the service provider's own staff did not carry out the majority of the training offered, customers perceived it to be the service provider's responsibility. The current position is not acceptable in the medium and long term, if the service provider wants to stand out with excellent service quality. The aim should be a higher perceived performance against this service standard. *Options:* (A) The service provider could try and take over those training elements currently performed by the subcontractor in question. *Pro:* Suitable staff are available within the service provider and could take over in the short run. *Contra:* There are likely to be certain obstacles as the subcontractor does not intend to give up those training activities and legal obligations may exist for the company to cease outsourcing the training. (B) The subcontractor's staff carrying out the training in question could be familiarized with the necessary customer processes and methods to be able to improve their performance. *Pro:* The performance is likely to be perceived better at least in the medium run. *Contra:* This would mean giving away core competences and further strengthening the subcontractor's position; that is, it would represent an investment in an outsider company instead of the company itself.

Subject	Analysis
(3) The *coaching quality* is very high.	This service standard was seen as very important, with the third highest priority amongst all service standards defined, and customers perceived the service provider to perform very well against this standard. As a consequence, its position lies well in the upper appropriate zone. Reasons for this seem to be that the staff carrying out coaching are highly motivated, very competent and helpful, and apparently manage to adjust their service delivery to meet their customers' needs very well. The position is very acceptable and does not need to be changed – it needs to be maintained. *Option:* The efforts within the service provider that have led to this good position need to be maintained and employees positively reinforced, in order to pre-empt complacency.
(4) The *helpdesk support quality* is very high.	The current position in the importance-performance matrix is just above the lower bound of acceptability in the appropriate zone, indicating that customers attach high importance to this service standard, the second highest priority after the speed of reaction to problems standard. This is consistent insofar as when customers have a problem they usually use the helpdesk support to get immediate help. Customers perceive performance to be reasonably good. Given the high importance, however, the position needs to be improved further up into the appropriate zone. *Options:* (A) The telephone numbers of the appropriate members of staff able to help with specific problems need to be communicated to customers. *Pro:* This helps to pre-empt customers from phoning the wrong people and being disappointed that they cannot be helped directly by them. *Contra:* For some customers it will be confusing to be given more than one or two numbers to choose from in case they have a problem. This may be compensated for by offering customers one 'general' contact who can then tell them who to phone in their specific case. (B) Helpdesk support staff (who are mostly the same people who also deliver coaching and training) could make sure that in case of non-availability of specific members of staff their telephone calls are automatically forwarded to the most suitable colleague. *Pro:* Customers calling are not handed around for a while or told to phone somebody else: they will in the best case not even notice that the member of staff who should normally answer the call is not available. *Contra:* There may be cases where it is not easy or even possible to identify the right person to forward the call to, if the person that is not available usually handles a combination of areas of expertise and cannot be replaced by just one colleague, or if there simply is no replacement available. In this case, calls should be forwarded to the general contact within the service provider.

Subject	Analysis
(5) *Promises* made are kept.	This service standard, although not of the highest customer priority, was still considered to be of importance to customers. However, the service provider was not seen to perform well against this service standard, so that the position in the matrix lies below the lower bound of acceptability in the improve zone. The reason for this seems to be the fact that a number of service promises that had been made by higher managers of the Support Services Group to customer management were not kept in the past. Hence, many customers were still sceptical about the dependability of the service provider's commitments. Clearly, the current position needs to be changed up into the appropriate zone. *Options:* (A) Members of the service operation should carefully watch what commitments they make (that is, what promises they make), both explicitly and implicitly. Those promises must be realistic ones only. *Pro:* The likelihood of not being able to deliver as promised is reduced if only realistic promises are made. *Contra:* There is still no guarantee that realistic promises can be kept and also customers may have the feeling that members of staff do not like to commit themselves. (B) Promises made could be actively followed up. *Pro:* This way it is less likely that promises are forgotten. *Contra:* There is an extra effort linked to following up promises. (C) Plan for effective recovery if promises cannot be kept; that is, it should be possible to turn around disappointed customers from the moment it becomes known that it is not feasible to deliver as promised. *Pro:* This way, negative word-of-mouth is pre-empted and disappointed customers can be turned around to even spread positive communications about the service operation's ability to deal with such difficult situations. *Contra:* There is an extra effort linked to such planning activities.
(6) The *overall impression* by support team members delivering services to the customer is very positive.	This service standard was perceived to be of medium importance only by most customers. Still, if employees of the service provider do not leave an overall positive impression (because they are unfriendly or seem arrogant or uninterested in the customers and their needs), the very basis of a long-term beneficial relationship with the service operation's customers is in peril. This is to say that, although customers attached medium priority to this standard, it must not be under-estimated. Also, due to its general and qualitative or subjective nature, performance measurements against this service standard can serve as an indicator of general perceived quality of the services provided. In the case study, customers perceived the service provider to perform reasonably well against this standard. The reasons for this outcome seem to be the fact that many customers had not had contact with most of the service provider's staff, and that some customers previously had contact with some members of the Support Services Group whom they

Subject	Analysis
	perceived to be arrogant – not really interested in customers' needs but rather in deploying 'their' specific tool (this was mentioned by different customers in individual discussions).
	Therefore, the current position in the matrix, although just in the appropriate zone, cannot be considered satisfactory. The position ought to be shifted further up the appropriate zone.
	Options:
	(A) Awareness of the importance of the overall impression could be increased amongst members of staff, using a relevant half-day or one-day seminar with all members of staff attending.
	Pro: There is a high probability that such a seminar would contribute very effectively to increased awareness of this issue amongst participating employees.
	Contra: The costs of such a seminar in terms of seminar fees and working hours of employees are substantial.
	(B) Awareness of the importance of the overall impression could be increased amongst members of staff, using an internal workshop with all members of staff attending.
	Pro: There is a high probability that such a workshop also would contribute very effectively to increased awareness of this issue amongst participating employees.
	Contra: Such a workshop would represent an investment in terms of working hours of participating employees.
	(C) Awareness of the importance of the overall impression could be increased amongst members of staff, using individual or group discussions of the issue with all members of staff.
	Pro: There is a high probability that individual and/or group discussions would contribute very effectively to increased awareness of this issue amongst members of staff.
	Contra: If some employees are stubborn and are not receptive to customer-centred approaches, individual discussions may not be sufficient to change those employees' behaviour towards customers.
(7) Support team members flexibly adjust to *working hours* and *locations* of the customers.	This service standard was perceived to be of the lowest priority to customers (along with the following standard). The service provider's performance against this standard was rated rather well though, so that the position in the matrix lies just at the border of the excess zone. This indicates that in light of the relatively low importance customers attached to this service standard it should be considered if efforts used to achieve this performance should or could not be used elsewhere (where it matters more to customers).
	The reason for this outcome seems to be that prior to this measurement, the general contact person of the service provider was almost the only member of staff seeing customer teams in one of the major sites covered, and he very flexibly adjusted his visits to customer schedules and locations, so that those customers concerned perceived delivery flexibility of the service operation to be very high.

Subject	Analysis
	This high degree of flexibility was possible also because private transportation was used between the sites covered (that is, extra costs were incurred privately), instead of using the company shuttle buses. Using those buses, however, would have significantly limited the service operation's flexibility to serve its customers. The position could be lower in the appropriate zone, considering that a lot of effort was needed to achieve this performance.

Options:

(A) Members of staff could – when needed – continue to use private transportation to cover both major sites and flexibly adjust to customers' working hours and locations.

Pro: This way the service operation would continue to perform high against this service standard.

Contra: There are financial, insurance, legal and private reasons that stand against the usage of private transportation. Also, shuttle buses are available, even if using them means that flexibility is reduced because they only allow to use a fixed time window at the other major site.

(B) Members of staff could use the shuttle services available between the major sites covered. Customers should be told, however, that those buses have to be used and, therefore, members of staff can only be on site within a fixed time window, in order to influence their expectations of the time flexibility of the services delivered.

Pro: No extra efforts for transportation needed, and customers do not expect members of staff to turn up outside the time windows pre-determined by the available shuttle services.

Contra: Lower perceived performance against this standard is to be expected, but this would be acceptable in light of the relatively low importance customers attach to it.

Subject	Analysis
(8) The service provider flexibly offers *new services* as the customer need arises.	This service standard was perceived (along with the previous one) to be of the lowest priority to customers. Still, the performance measured against this standard was quite poor so that the position in the matrix lies clearly below the lower bound of acceptability in the improve zone. The reason for this seems to be that many customers were not aware of the actual scope of services offered by the service provider; that is, they expected additional tools and processes to be supported by the service operation that were not part of the scope of services offered. Also the service provider could not easily make any such new tool or process part of its scope because it either did not have the resources or the know-how needed, or – which was often the case – because other units within the company were already charged with providing and supporting them. The current position is not acceptable and ought to change immediately up above the lower bound of acceptability into the appropriate zone.

Subject	Analysis
	Options:
	(A) The exact scope of services offered by the service provider (and which tools and processes lie outside this scope) could be clearly communicated to customers.
	Pro: Customers' expectations would be shifted into the right direction – they would be less likely to expect the service operation to deliver what the latter cannot.
	Contra: Some customers may feel that the service provider is trying to limit their scope of services where it gets difficult and then blame others as responsible for the delivery of services they are not capable of delivering properly.
	(B) Where possible and strategically of interest to the Support Services Group, the service provider could try and take over new elements into its scope of services.
	Pro: Customers would perceive the service provider's product/mix flexibility and its ability to form a long-term beneficial relationship with them to be high.
	Contra: If elements are taken on board that cannot be handled successfully for lack of relevant resources and/or know-how, then customers may lose confidence in the service provider.

done by different people. In the given case, only those options were presented that were considered to be most promising or at least seemed possible.

Other choices might well have led to similar or better results. This case study is not intended to offer to the reader the best solution in a specific case, but merely to give a practical example of an implementation of the service quality cycle.

After this analysis of the situation, Table 4.5 now shows the concrete action items derived from the previous analysis. Those action items were screened for compliance with the Support Services Group's strategy, as well as any constraints to do with contractual obligations, infrastructure, human resources and financial resources.

After the previous step, when concrete action items based on the analysis of the current situation were derived, the cycle now needed to be closed. Decisions had to be taken as to how and when those action items were best to be implemented and communicated, as well as anticipated resistance pre-empted. Also, it had to be decided which steps needed to be repeated and with what emphasis, when the cycle was to be gone through next time.

A longer meeting with all members of staff was scheduled as soon as possible, and all action items were communicated and discussed internally during this meeting. It was considered important that either all members of staff were convinced of the decisions taken and would support them, or that justifiable and constructive criticism actually was discussed during the meeting so that decisions could be altered into a better direction as early as possible.

The only action item that could not be internally implemented by the service provider was the one to take over all training from the one sub-contractor that was

Table 4.5   List of derived action items, March 2004

Subject concerned	Derived action items (concrete steps)
Situational context	All members of staff need to be aware of the situational context and of the negative attitude they may encounter when dealing with some of their customers, and of the underlying reasons for this attitude. Every service encounter must be used to indirectly convince customers of the usefulness of the services provided and that their negative attitude is not justified. Customers must be shown honest respect and appreciation.
(1) The *speed of reaction to problems* is very high.	Do nothing (= do not change anything).
(2) The *training quality* is very high.	Take over those training elements that are currently carried out by a subcontractor. Use the available human resources of the service provider who are able to perform this task satisfactorily and who have all the background knowledge necessary to better customize such training to customer needs.
(3) The *coaching quality* is very high.	Do nothing (= do not change anything).
(4) The *helpdesk support quality* is very high.	Communicate to customers which members of staff to contact concerning what specific subject (if they encounter a problem). Also make sure that calls are automatically forwarded to colleagues standing in for members of staff if the latter are not available. Where this is not possible, automatically forward calls to the overall contact person or point of the service operation.
(5) *Promises* made are kept.	Make sure that you and other members of staff only make explicit and implicit promises that are realistic, and follow them up until delivery as promised. Plan for effective recovery in case promises cannot be kept and turn around disappointed customers from the very moment you get to know that you will not be able to deliver to promises.
(6) The *overall impression* by support team members to the customer is very positive.	Increase awareness of the importance of this service standard amongst all members of staff, using individual and group discussions, such as during support team meetings that take place regularly anyway.
(7) Support team members flexibly adjust to *working hours* and *locations* of the customers.	All members of staff should use the shuttle buses available between the major sites to be covered. Customers should be told that the shuttle services have to be used, so that their expectations concerning the service provider's (delivery time) flexibility are shifted correspondingly.
(8) The service provider flexibly offers *new services* as the customer need arises.	Clearly communicate to all customers exactly what is and what is not in the scope of the services offered by the service provider, in order to set their expectations right. Where feasible and in the strategic interest of both customers and the Support Services Group, take new elements into the scope of services as the opportunity arises.

still used for one specific tool training. This issue had to be discussed as soon as possible and a decision triggered at higher levels of management within the Support Service Group, and possibly also within customer management.

Those action items of immediate relevance to customers had to be communicated externally as soon as the staff meeting was over and everybody in the service provider team had agreed on them.

Resistance, both internal and external, had to be pre-empted by convincing opinion leaders (in the case of the service provider every staff member was considered an opinion leader because they all had direct contact with customers), not just by imposing the changes implied.

Communication to customers was by means of reporting by the general focal point (Capability Integrator) that took place on a weekly basis to all customers (formal), as well as during personal encounters between members of staff and customers (informal).

Second Cycle Run

The next time the cycle was gone through, emphasis was decided to be laid on Steps 3, 4 and 5 of the method; that is, the performance measurement, the analysis of the current situation and the derivation of action items. The reason for this was that both the definition and ranking of service standards seemed to be stable over time. And since customers should not be annoyed with measurements if they do not seem

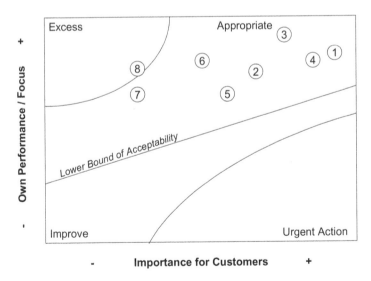

Figure 4.6   Performance against service standards, May 2004

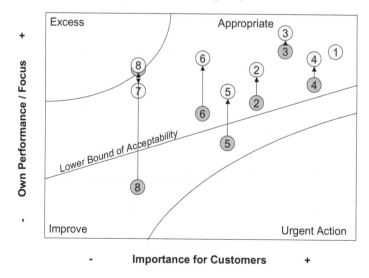

Figure 4.7   Tendencies, March–May 2004

necessary (to minimize the impact on their work), it seemed a good idea not to go through every step of the method again every time the service provider goes through the cycle.

The next step after closing the cycle was the next performance measurement in May 2004, similar to the one undertaken when the service quality cycle was first implemented. The only differences were that the number of customers contacted had slightly risen and, as was decided above, the questionnaire used – this time in English and German – was sent out as part of an e-mail, not an attachment. Figure 4.6 presents the results of the May 2004 measurements, using the same numbering symbols for the service standards as before.

Figure 4.7 shows the perceived performance improvement or decrease: the present position of each service standard is indicated by a lightly shaded circle, as opposed to the previous position, indicated by darker shading.

Once again, it should be stated that this was not necessarily a statistically clean exercise, because the participants were not necessarily the same customers in both measurements, nor was the number of participants very high, and also a mix of means was used to question customers. But the essential thing here was to merely get an indication of where the service provider seemed to be and where it seemed to be going.

Table 4.6 covers the discussion and analysis of each of the service standards as well as the current situational context in May 2004, in order to form a sound basis for operational and strategic decision-making in the form of concrete action items later on.

Table 4.6   Analysis of the current situation, May 2004

Subject	Analysis
Situational context	Most of what was summarized in Table 4.4 is still valid, although many customers now think more positively about the service provider. Most new customers show more positive attitudes, because they are now delivered support services, in contrast to the period when they had not been or only been poorly supported by the service provider from France (for the reasons mentioned above). Those tool training elements that were performed by external resources before the last measurements have now all been taken over by members of staff of the service provider.
(1) The *speed of reaction to problems* is very high.	Unchanged position (see Table 4.4). *Option:* The efforts within the service provider that have led to this good position need to be maintained and employees positively reinforced, in order to pre-empt complacency.
(2) The *training quality* is very high.	The perceived performance against this service standard has improved significantly since the last measurement. The reason for this seems to be the take-over of the tool training (previously outsourced) by very suitable own human resources of the service provider. The current position clearly lies in the appropriate zone, indicating that current performance is satisfactory, considering the customer-perceived importance of this service standard. No change of position is needed. *Option:* Maintain current efforts and positively reinforce employees.
(3) The *coaching quality* is very high.	The performance against this service standard was rated even better than in the last measurement, moving the position further into the upper appropriate zone. The reason for this seems to be that more customers have had the opportunity to experience the coaching offered which is still delivered at the same high quality as before. No change of position is needed. *Option:* Maintain current efforts and positively reinforce employees.
(4) The *helpdesk support quality* is very high.	Performance against this service standard is rated significantly higher than in the previous measurement so that the position now lies well in the upper appropriate zone. No change of position needed. *Option:* Maintain current efforts and positively reinforce employees.

Subject	Analysis
(5) *Promises* made are kept.	Significant improvement of perceived performance against this service standard so that position now lies well above the lower bound of acceptability in the appropriate zone. No change of position needed. *Option:* Maintain current efforts and positively reinforce employees.
(6) The *overall impression* by support team members delivering services to the customer is very positive.	Significant improvement of perceived performance against this service standard so that position now lies much further up in the appropriate zone. One reason for this also seems to be the fact that more customers have had contact with more or all members of staff than was the case before the previous measurement. No change of position needed. *Option:* Maintain current efforts and positively reinforce employees.
(7) Support team members flexibly adjust to *working hours* and *locations* of the customers.	The perceived performance against this standard was lower than in the previous measurement so that the position now lies well in the appropriate zone. The reason for this seems to be the fact that all members of staff have been using the shuttle buses available to travel between sites, so that delivery time flexibility of the services offered was reduced to a certain degree. The current position is very much acceptable. No change of position needed. *Option:* Maintain current efforts and positively reinforce employees.
(8) The service provider flexibly offers *new services* as the customer need arises.	Performance against this service standard was very high so that the position lies just in the excess zone. The reason for this seems to be the shift in customers' expectations of what the scope of services of the service provider actually is, thanks to communicating the scope of the services offered to all customers since the last measurement. Customers now perceive the service provider to be actually quite flexible concerning new services. The position is very much acceptable, although attention should be paid to not spending too much extra effort on achieving this position. No change of position is needed. *Option:* Maintain current efforts and positively reinforce employees.

After this analysis of the situation, Table 4.7 now shows the concrete action items derived from the previous analysis. Those action items were screened for compliance with the Support Services Group's strategy, as well as any constraints to do with contractual obligations, human resources and financial resources.

Table 4.7   List of derived action items, May 2004

Subject concerned	Derived action items (concrete steps)
Situational context	All members of staff need to be aware of the situational context and of the negative attitude they may encounter when dealing with some of the customers, as well as with the underlying reasons for this attitude. Every service encounter must be used to indirectly convince customers of the usefulness of the services provided and that their negative attitude is not justified. Customers must be shown honest respect and appreciation.
(1) The *speed of reaction to problems* is very high.	Do nothing (= do not change anything).
(2) The *training quality* is very high.	Do nothing (= do not change anything).
(3) The *coaching quality* is very high.	Do nothing (= do not change anything).
(4) The *helpdesk support quality* is very high.	Do nothing (= do not change anything).
(5) *Promises* made are kept.	Do nothing (= do not change anything).
(6) The *overall impression* by support team members delivering services to the customer is very positive.	Do nothing (= do not change anything).
(7) Support team members flexibly adjust to *working hours* and *locations* of the customers.	Do nothing (= do not change anything).
(8) The service provider flexibly offers *new services* as the customer need arises.	Do nothing (= do not change anything).

In this case, the outcome was that nothing should be changed; in other words, the current efforts were apparently going into the right direction.

Having completed this step, it was decided – based on the analysis of the current situation – that no changes seemed to be necessary and that the cycle now needed to be closed again; that is, decisions had to be taken as to which steps of the cycle needed to be repeated (and with what emphasis) when the cycle was gone through next time (see discussion above after the March 2004 derivation of action items in Table 4.5).

Third Cycle Run

The next time the cycle was gone through, emphasis was decided to be laid – once again – on Steps 3, 4 and 5 of the method; that is, the performance measurement, the analysis of the current situation and the derivation of action items. As before,

Delivering Excellent Service Quality in Aviation

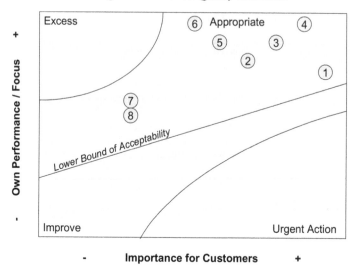

Figure 4.8	**Performance against service standards, July 2004**

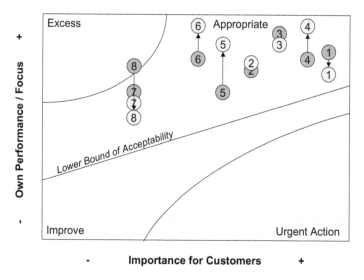

Figure 4.9	**Tendencies, May–July 2004**

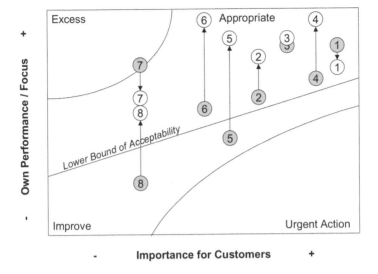

Figure 4.10   Tendencies, March–July 2004

the reason for this was that both the definition and ranking of service standards seemed to be stable over time. And since customers should not be annoyed with measurements if it does not seem necessary (to minimize the impact on their work), it was not considered sensible to go through every single step of the method again.

The next step after closing the cycle was the performance measurement in July 2004, similar to the one undertaken before. Figure 4.8 shows the results of the July 2004 measurements, using the same numbering symbols for the service standards as before.

Figures 4.9 and 4.10 show the tendencies compared with the May 2004 and the March 2004 values respectively, in order to give an indication of perceived performance improvement or decrease over these time intervals.

Table 4.8 covers the discussion and analysis of each of the service standards, as well as the current situational context, in order to form a sound basis for operational and strategic decision-making in the form of concrete action items later on.

After this analysis of the situation, Table 4.9 shows the concrete action items derived from the previous analysis. Those action items were screened for compliance with the Support Services Group's strategy, as well as any constraints to do with contractual obligations, human resources and financial resources.

Having gone through the service quality cycle three times so far in the course of this case study, with special emphasis on Steps 3, 4 and 5 (once the service standards and their relative importance to customers had been established and seemed to be stable over time), it was decided to go through the cycle again after a longer time interval of about four to five months instead of the two-month intervals used so far,

Table 4.8   Analysis of the current situation, July 2004

Subject	Analysis
Situational context	What was said in Table 4.6 is still valid.
	However, due to a budget issue outside the reach of the service provider, crucial resources that were needed to administer one of the most important tools used by the service provider's customer teams were no longer financed and had to leave the company.
	This critically hampered customers' daily work and seriously put in peril one overall programme milestone – the delivery of specific documents in time and at the right quality.
	Although it was immediately communicated to customers that the problem was not the responsibility of the service provider, a negative climate was developing and members of staff received many complaints.
(1) The *speed of reaction to problems* is very high.	The current position in the importance-performance matrix is in the appropriate zone and indicates that this service standard is perceived to be the most important one of the entire set of service standards.
	However, compared to the last measurements, customers perceive the service provider to perform slightly worse but still well against this standard.
	The reason for this slightly worse perception is likely to be that, although customer contact staff of the service provider are still highly motivated, the lack of specific resources has led to the impression that not all customer problems are resolved immediately. Specifically this was the case for those that could only be resolved by the very resources that had not been available in this last period.
	Although this position could still be considered to be all right, efforts should be made to regain the previous position.
	Options:
	(A) The scarce resources that are urgently needed to resolve those customer problems currently unresolved could be re-employed.
	Pro: Instant recovery of the situation.
	Contra: Budget issues that had led to the unavailability of the resources in question are still not likely to be resolved so that this option could on its own only be a short term approach.
	(B) The underlying budget issues need to be resolved.
	Pro: Medium- or long-term solution of the problem.
	Contra: Clarification of those issues may take a while so there would be no short term recovery. Also, the resources needed are not likely to be still available for the company, if it takes too long until they are re-employed.
	(C) Existing staff could be trained to do the work of the personnel who are currently not available.
	Pro: This may be a possible medium- to long-term solution, which may or may not motivate the staff concerned, depending on how heavy they perceive their current workload to be.
	Contra: Staff concerned may not be motivated to take over this additional work. Also, there is no short-term recovery of the situation.

The staff concerned may not be suitable or even available for the tasks to be taken over.

(D) If it is outside the influence of the service provider to resolve the underlying budget issues, this very fact could be communicated to the customers.

Pro: Customers may be able and willing to increase the pressure on those who can actually influence the resolution of the underlying budget issues. Also, customer expectations from the services delivered (or in this case not delivered) to them would be altered in a way that they do not perceive the service operation's performance to be poor because they know that the current problem is out of the reach of the service operation.

Contra: Some customers may think the service provider is trying to excuse its way out of the situation, although it really is the latter's responsibility to resolve the current problem.

(2) The *training quality* is very high.	The position has even further improved compared to the last measurement (see Table 4.6). No change of position needed.

Option:
Maintain current efforts and positively reinforce employees.

(3) The *coaching quality* is very high. The position is still high up in the appropriate zone, although slightly lower than in the previous measurement, indicating that customers perceive coaching quality to be very high.

The slight decrease in measured performance should not be over-interpreted as a problem.
No change of position needed.

Option:
Maintain current efforts and positively reinforce employees.

(4) The *helpdesk support quality* is very high. Customer-perceived performance against this service standard has even further improved since the last measurement so that the position now lies at the top of the appropriate zone.
Excellent position, no change of position needed.

Option:
Maintain current efforts and positively reinforce employees.

(5) *Promises* made are kept. Even further improvement of perceived performance against this service standard so that position now lies in the upper appropriate zone.
No change of position needed.

Option:
Maintain current efforts and positively reinforce employees.

(6) The *overall impression* by support team members delivering services to the customer is very positive. Even further improvement of perceived performance against this service standard so that the position now lies at the top of the appropriate zone.
Customers perceive members of staff as very friendly, customer oriented, not at all arrogant, as well as very pleasant to work with.
No change of position needed.

Option:
Maintain current efforts and positively reinforce employees.

(7) Support team members flexibly adjust to *working hours* and *locations* of the customers.	Measured performance against this service standard was slightly lower than in the previous measurement, but not significantly. The current position lies well in the appropriate zone. No change of position needed.
	Option: Maintain current efforts and positively reinforce employees.
(8) The service provider flexibly offers *new services* as the customer need arises.	Performance against this service standard was clearly lower than in the last measurement, but the position now still lies well in the appropriate zone. The reason for this seems to be that customer expectations from the scope of services offered are now more realistic, after initially being too high and then too low. The position is very acceptable. No change of position needed.
	Option: Maintain current efforts and positively reinforce employees.

since the beginning of the case study. In doing this, Step 1 should be left out and the customer involvement parts of Steps 2 and 3 combined. The reason for this was that, first, the summer holiday period and the relatively good position across all service standards seemed to allow for a longer interval until the next measurements. Second, customers should not be annoyed with too many interviews and questionnaires in too short a time interval unless there is a good reason for this (for example, if there is an urgency to keep track on short-term developments of the service operation's position against specific service standards). Third, it was felt that the priority of the service standards should be questioned or challenged again, combined with the next measurement of performance against the service standards, so that any potential changes in the perceived relative importance of the service standards could be detected.

In addition to that, it was decided to split Service Standard 5, 'Promises made are kept', into two parts for the next performance measurement, one related to the offered services (which could be directly influenced by the service provider) and one linked to new tools, versions, updates, add-ons, etc. (which could not be directly influenced by the service provider). The reason for this change was the wish to be able to measure the difference between how customers perceived the service provider as keeping service promises and how reliable they perceived statements to be that were related to the arrival of new tools, etc. The latter could only be influenced indirectly by the service provider by means of realistic and careful communication of such statements.

Fourth Cycle Run

The next time the cycle was gone through, emphasis was decided to be laid – once again – on Steps 3, 4 and 5 of the method (that is, the performance measurement, the analysis of the current situation and the derivation of action items. However, since

Table 4.9   List of derived action items, July 2004

Subject concerned	Derived action items (concrete steps)
Situational context	All members of staff need to be aware of the situational context and of the negative attitude they may encounter when dealing with some of the customers, as well as the underlying reasons for this attitude. Every service encounter must be used to indirectly convince customers of the usefulness of the services provided and that their negative attitude is not justified. Customers must be shown honest respect and appreciation. Also, make it clear to all customers that you do all you can both to put pressure on those people who can resolve the current budget issues mentioned in Table 4.8 and to improvise and help customers ease the resulting situation as far as possible. As soon as angry customers directly offend you, take their concerns very seriously but clearly state where your limits of responsibility are, always treating them with respect.
(1) The *speed of reaction to problems* is very high.	Communicate the problem to all stakeholders in the service operation including all customers, in order to both shift expectations and increase the pressure to resolve the underlying budget issues. In the short term, try and re-employ the resources currently not available for short-term recovery of the situation. In doing so, be careful not to ease the pressure on those who can resolve the underlying budget issues.
(2) The *training quality* is very high.	Do nothing (= do not change anything).
(3) The *coaching quality* is very high.	Do nothing (= do not change anything).
(4) The *helpdesk support quality* is very high.	Do nothing (= do not change anything).
(5) *Promises* made are kept.	Do nothing (= do not change anything).
(6) The *overall impression* by support team members delivering services to the customer is very positive.	Do nothing (= do not change anything).
(7) Support team members flexibly adjust to *working hours* and *locations* of the customers.	Do nothing (= do not change anything).
(8) The service provider flexibly offers *new services* as the customer need arises.	Do nothing (= do not change anything).

the relative perceived importance of the service standards seemed to have somewhat changed, Step 2 of the service quality cycle (the ranking of service standards) was decided to be combined with the performance measurement (to minimize the impact on customers' work). In the questionnaire, customers were asked to both rank and rate per service standard.

Depending on where an aircraft development programme stands in the overall development process, different aspects of service delivery appeared to be of importance. Also, there seemed to be an increasing gap between the perceived dependability of the service operations insofar as service promises that were related to the pure delivery of services by members of the service provider seemed to be quite high, whereas promises made in connection with the delivery of new tool versions, add-on tools, etc. (that could not be directly influenced by the service provider) could often not be kept. Therefore, although the standard 'Promises made are kept' was decided to be maintained, performance against this standard was to be measured taking into account this assumed gap. Customers were to be asked how they perceive the service provider to perform against this standard concerning their own services and also related to new tool versions, updates, add-ons, etc., that could not be directly influenced by the service provider.

Figure 4.11 shows the results of the December 2004 measurements, using the same numbering symbols for the service standards as before.

Figures 4.12 and 4.13 show the tendencies compared with the March 2004 values in order to give an indication of perceived performance improvement or decrease

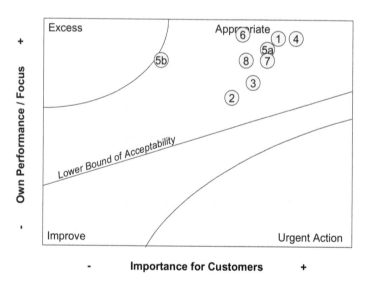

Figure 4.11   Performance against service standards, December 2004

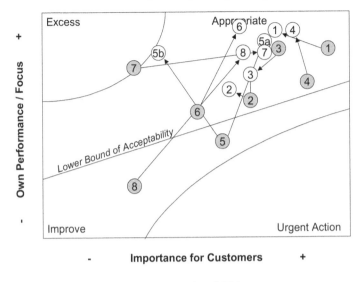

Figure 4.12 Tendencies, March–December 2004

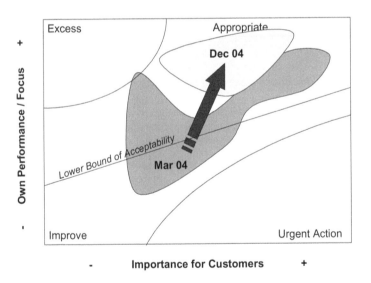

Figure 4.13 Tendencies, March–December 2004 (simplified)

over the entire time interval from March to December 2004. Figure 4.12 shows all individual service standards as for the previous measurements, whereas Figure 4.13 shows a simplified view of the performance development across all service standards in the form of calculated areas that represent the distribution of the corresponding service standards in the diagram without showing the position of each individual

Table 4.10   Analysis of the current situation, December 2004

Subject	Analysis
Situational context	The general climate overall was positive and constructive, and all new team members of the service provider had settled well into the team. They were by now well familiar with their customers and the services they were expected to deliver to them.
	Many additional customer teams from three ACMTs (as opposed to one at the beginning of the year) had started using the services offered and new services were flexibly being developed and delivered to specific customer teams.
	Service agreement letters were prepared between each of those ACMTs and the service provider. Those contracts were to be based on previous demand analyses at CMIT level and fixed internal hourly rates.
	Based on these new figures, the contracts to buy in external resources to staff the service provider were prepared (by now about 80 per cent of the resources used to deliver the services were external specialists).
	Most customers were under tremendous time pressure because an important aircraft programme milestone was to be passed in December. This apparently resulted in two things: first, customers were being very demanding and depending on the service provider's activities; second, many customers would not take their time to give the service provider feedback on their performance so that the reply rate to the questionnaire was considerably lower than was the case in the previous performance measurements.
(1) The *speed of reaction to problems* is very high.	The current position in the importance-performance matrix is well within the appropriate zone and indicates that this service standard is still perceived to be important, although no longer the most important one of the entire set of service standards. Compared to the last measurements, customers perceive the service provider to perform clearly better against this standard.
	The reason for this is likely to be that the lack of specific resources prevalent at the last performance measurement had been overcome since then and the service provider was now sufficiently staffed with skilled personnel to allow for short response times to any problems. Excellent position, no change of position needed.
	Option: Maintain current efforts and positively reinforce employees.

Subject	Analysis
(2) The *training quality* is very high.	The position lies in the appropriate zone but perceived performance of the service provider has decreased significantly. Also, customers consider this service standard as being slightly less important than before. The reason for the former seems to be a change in how customers can get registered for training in one of the core capabilities offered by the service provider. The administration of the training was now performed by some other department (planning, scheduling, inviting, billing, etc.), whereas the same resource from the service provider still actually delivered the training. Before the change, customers could choose from several training dates online and could also see where training places were still available and how many. They could then register themselves and get immediate confirmation. After the change, customers had to send their name to the new department in charge of the training's administration: this department would collect all names of interested people, decide on a date later and invite the first batch of them to the training session. As expected, a large percentage of the first batch could not actually make it on that date, so that often training sessions were cancelled or took place with a ridiculously small number of participants. The results were that many customers were angry about the new system and the training cost per customer was rising. *Options:* (A) The new department in charge of the administration of the training could be brought to change their registration, planning and invitation procedure. *Pro:* Customers would have the same advantages they had before the new department took over the administration of trainings; that is, concrete dates to choose from and immediate confirmation of their registration. *Contra:* The new department apparently used a software to administer the trainings that did not allow for this change back to the previous situation. Also, their procedure was applied by the department transnationally, which would make a change even harder.) (B) The user needs and demand for training could be centralized by the service provider who could then negotiate training dates with the new department in charge. *Pro:* This approach would make life easier for customers with all the advantages mentioned above. *Contra:* It would also mean a considerable increase of the workload of some members of the service provider who would have to act as an intermediate 'negotiator' between customers and the department. (C) Leave the situation as it is, communicate the reasons for the problems to customers and emphasize that it is out of the service provider's reach to change the situation. *Pro:* This approach would appear to be the easiest way, at least in the short run. No extra workload would be created for the service provider.

Subject	Analysis
	Contra: Customers would not really be helped. From the overall company perspective, resources would be wasted because training sessions would have to be scheduled, cancelled and rescheduled on a regular basis, and the average number of participants per training would be lower than could be possible, resulting in higher training costs per person.
(3) The *coaching quality* is very high.	The position lies still in the appropriate zone although significantly lower than in the previous measurement, indicating that customers perceive coaching quality to be a bit lower. Also, the perceived importance of this service standard has decreased slightly. The former could have been caused by time pressure under which those team members delivering the coaching had to work in light of the imminent programme milestone, and maybe the duration of the coaching session was not communicated clearly enough to customers beforehand. The latter may be caused by the fact that the most important capability at this stage of the aircraft development process was already deployed in all customer teams and many customers would rather use hotline than coaching services concerning this capability.
	Options: (A) Focus on concrete customer needs, devote sufficient time to dealing with few but relevant customer issues and tackle 'normal' coaching needs with training and hotline services. *Pro:* The expertise and skills of individual members of the service provider could be used to great benefit in order to solve specific issues that are critical to certain customer teams. More general coaching needs could be addressed by the team's training and hotline services. *Contra:* Customers may be in a situation where those general coaching needs cannot be met immediately or in the short run by training services due to the problems mentioned above that are linked to the current training registration procedure. (B) Leave the situation as it is, do not change anything. *Pro:* Seemingly the simplest way ahead (at least in the short run) with no major resistance to be expected. *Contra:* If the service provider intends to keep being involved in the more strategic and critical issues of the development programme, its staff have to concentrate their coaching efforts on participation in conceptual work with different customer teams and corresponding interface partners as opposed to coaching normal or standard training knowledge that could also be transferred to customers by training or hotline (in specific cases).
(4) The *helpdesk support quality* is very high.	The position still lies very much at the top end of the appropriate zone, with only marginal difference to the previous measurement result. Also, the perceived importance of this service standard has not significantly changed. Excellent position, no change of position needed.
	Option: Maintain current efforts and positively reinforce employees.

Subject	Analysis
(5) *Promises* made are kept.	Performance against this standard was measured in two directions: first, concerning the own services provided such as hotline services or the import of a number of documents into another tool that could directly be influenced by the service provider; second, concerning the arrival of new versions or updates of tools or add-ons, which was communicated by but could not directly be influenced by the service provider. The results of the measurement showed that customers perceived the service provider to be very reliable concerning their own services and also that this was considered to be of high importance to customers. Since there had been some delays in the delivery of new versions of tools and add-ons, it could have been expected that customers perceived the service provider to be somewhat less reliable from this viewpoint. Maybe surprisingly, customers rated the service providers performance to be very high even concerning those aspects of the service standard they could not directly influence. However, customers attached far less priority to this than to the reliability concerning their own services. The reason for this seemed to be the fact that members of the service provider made very cautious statements and promises about the schedule of arrivals of new tools or add-ons, so as to communicate to customers the given degree of uncertainty and that they could not directly influence the reliability of the suppliers of that software. No change of positions needed. *Option:* Maintain current efforts and positively reinforce employees.
(6) The *overall impression* by support team members delivering services to the customer is very positive.	The position still lies at the top of the appropriate zone. Customers perceive members of staff as very friendly, customer-oriented, not at all arrogant, as well as very pleasant to work with. The perceived importance of this service standard has even increased. Very good position, no change of position needed. *Option:* Maintain current efforts and positively reinforce employees.
(7) Support team members flexibly adjust to *working hours* and *locations* of the customers.	The current position lies well in the appropriate zone. Customers perceived the service provider to perform very well against this service standard and at the same time it was now perceived to be of considerably higher importance than had previously been the case. Good position, no change of position needed. *Option:* Maintain current efforts and positively reinforce employees.
(8) The service provider flexibly offers *new services* as the customer need arises.	The position lies well within the appropriate zone. The service provider was perceived to perform very well against this service standard and at the same time customers now perceived it to be of considerably higher importance than had previously been the case. Good position, no change of position needed. *Option:* Maintain current efforts and positively reinforce employees.

standard. This helps to give a general indication, especially to people who are not familiar with the details of specific service quality management, but must be taken with care and an analysis of the more detailed position of each service standard is still indispensable for decision-making.

In the case study, Figure 4.13 shows a clear reduction of the distribution area and a general trend upwards towards better-perceived performance and to the right part of the diagram; that is, towards higher perceived importance (on average). This can be argued as indicating that, in general, customers now seemed less differentiating concerning the relative importance of each service standard and that they perceived the service provider to clearly perform better against their service standards.

Table 4.10 covers the discussion and analysis of each of the service standards as well as the current situational context, in order to form a sound basis for operational and strategic decision-making in the form of concrete action items later on.

After this analysis of the situation, Table 4.11 now shows the concrete action items derived from the previous analysis. Those action items were screened for compliance with the Support Services Group's strategy, as well as any constraints to do with contractual obligations, human resources and financial resources.

Table 4.11   List of derived action items, December 2004

Subject concerned	Derived action items (concrete steps)
Situational context	All members of staff need to be aware of the situational context and of the time pressure under which customers currently have to work.
	Customers must rightfully have the impression that all team members are aware of the current situation and the prevalent urgency.
(1) The *speed of reaction to problems* is very high.	Do nothing (= do not change anything). Specifically keep: – being highly available (by phone and e-mail) – using mobile phones for hotline services (also during internal meetings) – using the message function and calling customers back when they have tried in vain to reach you – making clear replacement agreements – forwarding e-mails internally without extra text if no explanation is needed and the message went to the wrong team member. – using a customer database where feasible.
(2) The *training quality* is very high.	Leave the situation as it is, communicate the reasons for the problems to customers and emphasize that it is out of the service provider's reach to change the situation. However, keep reporting and communicating in other ways and to the right people who could actually improve things within the new department, what advantages the previous registration procedure had and how customers perceived the difference between both approaches.

Subject concerned	Derived action items (concrete steps)
	Concerning the training itself, do not change anything. Specifically, keep: – communicating agenda and objectives beforehand – adjusting training material to general user needs – identifying individual needs and interests – answering individual questions – using real-life exercises – offering black and white handouts – ordering drinks for attendees.
(3) The *coaching quality* is very high.	Focus on concrete customer needs, devote sufficient time on dealing with few but relevant (strategic) customer issues and tackle 'normal' coaching needs with training and hotline services. Concerning the coaching services offered keep: – listening to customers – making sure the relevant members of the service provider have sufficient time – communicating realistic durations of coaching services to customers – focusing on concrete needs of customers.
(4) The *helpdesk support quality* is very high.	Do nothing (= do not change anything). Specifically, keep: – being highly available – using mobile phones – also answering calls during internal meetings.
(5) *Promises* made are kept.	Do nothing (= do not change anything). Specifically, keep: – making realistic promises (rather under-promise) – following up (e.g. using a customer database) – communicating if it is not your fault that a promise could not be kept – using customers to build up pressure on those responsible for failure to deliver as promised – over-delivering.
(6) The *overall impression* by support team members delivering services to the customer is very positive.	Do nothing (= do not change anything). Specifically, keep: – using customer language – showing customers honest respect and appreciation as persons, consumers of your services and experts in their respective fields – making an up-to-date chart of your team available (including 'nice' photos and telephone numbers under which customers can reach each member of the team).
(7) Support team members flexibly adjust to *working hours* and *locations* of the customers.	Do nothing (= do not change anything). Specifically, keep: – seeing customers at their convenience (within reason, such as time/location limits) – using shuttle buses (that are provided without extra cost) whenever possible – communicating limitations to customers (for example, that shuttle buses are to be used and you are dependant on the bus schedules).

Subject concerned	Derived action items (concrete steps)
(8) The service provider flexibly offers *new services* as the customer need arises.	Do nothing (= do not change anything). Specifically, keep: – identifying future customer needs – taking first steps without delay (trigger investigations into solutions to meet new customer needs) – keeping customers informed of the status of specific investigations – explaining the reasons if certain investigations are not followed up.

Having gone through the service quality cycle four times so far in the course of this case study, with special emphasis on Steps 3, 4 and 5, and leaving increasing time intervals between measurements, it was decided to go through the cycle again after a longer time interval of about five to six months, after the next programme milestone. The reason for this was that, first, the relatively good position across all service standards seemed to allow for a longer interval until the next measurements and, second, customers should not be annoyed with too many interviews and questionnaires in too short a time interval unless there is a good reason for this (such as if there is an urgency to keep track on short-term developments of the service operation's position against specific service standards).

Step 1 was to be left out and the customer involvement parts of Steps 2 and 3 combined again to take into account the fact that during the overall aircraft development process the perceived priorities of individual service standards seemed to change.

After looking at the implementation of the service quality cycle in this section, the next section describes the status after the implementation, as opposed to the status before the implementation that was described in the previous section.

Status After

The service provider now carried out local support for all customer teams in Germany. However, since different customer teams had different needs because they were at slightly different stages in their development process or they carried out different tasks in the overall aircraft development programme, the service provider supported them differently or individually depending on their specific needs.

Although all (currently identified) potential customer teams were now served, some teams were only served in one or two capabilities (supporting the specific methods and tools used by those teams), others were served across a wider range of support services. Some teams used one capability to develop and specify concrete pieces of equipment, others used the same capability to specify transversal requirements that had to be applied by all other development teams.

Most customer team members had joined their respective teams by now and at least gained a few months of specific work experience in the aircraft development programme under consideration.

Time pressure on customer teams was still quite high, with engineering deliverables such as requirements and design specifications having to be produced in time and in the right quality at specific programme milestones.

In the meantime, the resources that had been allocated from the relevant Capability teams to the Program team in Germany (that is, the service provider) were usually sufficient to cover the spectrum of services offered. However, the budget situation to finance those resources was a bit tricky to manage, because of the long lead times from the generation of budget input figures via the negotiation of the budget allocated to actually making the budget available to the service provider. In other words, new service support needs of customers that had not previously been taken into account (for yearly planning) had to be compensated for – in the short and medium term – by cutting or streamlining other activities of the service provider.

With respect to proactive service quality management, there were now customer-driven service quality standards defined and communicated, both internally to employees and externally to customers. Also, customer-perceived performance against those standards had been measured in two- to five-month time intervals, to enable systematic improvement and control of service quality.

The first positive effects of this approach had become obvious and clearly contributed to boost the motivation of all members of staff, and enable better decisions, both (day-to-day) operationally and strategically.

The reputation of the Support Services Group amongst customers in Germany that were now served by the service provider had improved significantly. Two good indicators for this can be argued to be the perceived performance ratings of the service provider against service standards 5 and 6 over the 12-month period under consideration ('Promises made are kept' and 'The overall impression by support team members delivering services to the customer is very positive').

Figure 4.14 gives a simplified overview of the coverage of specific capabilities (that is, the range of support services offered) over the percentage of potential customers that are addressed in Germany, as at the end of this case study. As can be seen in Figure 4.14, at the end of the present case study all (currently identified) customer teams were now covered and a wider range of support services or capabilities offered to customers by the service provider.

For the reasons mentioned in the beginning of this section, customer teams were supported individually, depending on their different needs, by covering only those capabilities relevant or currently needed by them, given their specific circumstances in the overall aircraft development process.

Figure 4.15 shows the actual and estimated (from year 2006) budget development from 2004 till 2008 for the service provider, concerning one specific aircraft development programme. The required budget, as estimated and requested by the service provider, is contrasted with the customer budget estimations over the same period.

Tendencies in all budget lines are similar but at quite a different order of magnitude. They show the usual stages of a development cycle. There is a clear ramp up in the beginning of the programme in 2004, followed by a stagnation point in 2005, a rapid decrease in 2006 and, last, a slow phase-out period until 2008.

The agreed budget line shows the result of tough negotiations between the service provider and different customer organizations that are part of the aircraft development

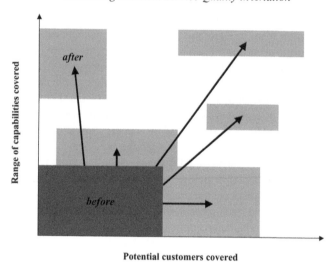

Figure 4.14   Support services coverage in Germany (status after)

(Estimated from 2006)

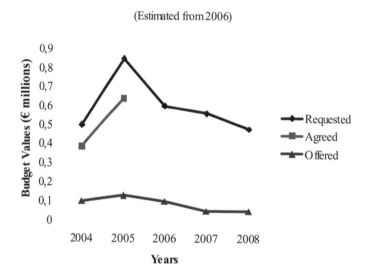

Figure 4.15   Budget development over time

programme under consideration. During such negotiations, it is essential to be able to convince customers that the services offered are worthwhile paying for, both because they are necessary or beneficial for the business and because they are of very high quality. Anything that helps to visualize how beneficial the services are to the customer and that the services delivered are of very high quality is most valuable during such discussions.

The results of service quality measurements as proposed in this book are, therefore, most useful to demonstrate to customers that the service provider is very much aware of what customers expect from its services and how customers perceive the provider's performance. In other words, the outcomes of the proposed method not only help to improve service quality as such, but they are also highly effective in convincing customers that high quality services are being delivered. This, in turn, directly supports contract negotiations between the two parties and contributes to customers' willingness to give the service provider the budget needed.

The service provider or its individual team members are partly deployed in different aircraft development programmes that are simultaneously at different stages of the development process. This allows for both continuous availability of the highly skilled resources needed and beneficial exploitation of synergies between different aircraft programmes. During 2004, mainly three persons worked for the service provider, although not all for 100 per cent of their working time. In 2005, the team consisted of up to ten members, most of whom did not work full-time for the aircraft programme under consideration.

Shortly after the present case study, during a strategic workshop of the Support Services Group in France, the results of this case study were presented and discussed. It was decided to prepare the extension of the use of the method proposed in this book to all other service providers of the group per country and aircraft programme by the end of June 2005, with the actual transnational implementation to follow from July 2005.

In order to facilitate and prepare the implementation of proactive service quality management along the lines suggested, a small transnational working group was formed and mandated to locally coach and enable the leaders of all support service teams to implement the service quality cycle for their own service operations.

The previous sections have described the present case study, the situation before, the implementation of the service quality cycle and the situation afterwards: the following section will summarize the major lessons that can be learned from this case study.

Lessons Learned

Table 4.12 summarizes major lessons learned from the present case study, categorized by specific and related topics. Overall, it can be said that the application of the service quality cycle as proposed in Chapter 3 has proven to be:

Table 4.12   Lessons learned

Topics	Lessons learned
Customer contact	• Regular personal contact with customers by members of staff and representation of the service provider in relevant customer meetings were perceived to be most important in order to gain valuable information to be able to better provide the services offered (such as customer schedules, complaints, problems, etc.), to communicate to customers (own constraints, schedules, next steps, service standards, etc.) and to generally reassure customers of the services provided (compensation for the lack of tangibility of the services offered). • Showing honest respect and appreciation in dealing with customers has been proven to give members of staff the right attitude when delivering the services offered. Also, customers rightfully feel taken seriously and are shown the respect they deserve as people, consumers of the services delivered and the experts they are in their specific fields of activities. It is important that this respect and appreciation is meant seriously (out of conviction). However, this does not preclude that sometimes unrealistic demands from customers have to be refused, but if so in a polite manner that corresponds to the given situation.
Flexibility of the service quality cycle	• The service quality cycle has been shown to be a very flexible method, that can easily be adjusted to a given situation in terms of what steps are used per cycle or which is their relative emphasis in a given cycle. In this case study, for instance, it was found that the service standards and their customer-perceived importance were very stable, so that in the second and third cycle the corresponding steps of the method (Steps 1 and 2) could be left out. However, during the overall aircraft development process, perceived priorities do seem to change over time so that from time to time at least Step 2 should be included, even if combined with Step 3 to reduce the impact on customers.
Combining customer involvement parts of several steps of the service quality cycle	• Several sub-steps related to the customer involvement parts of Steps 1, 2 and 3 can be combined to save own efforts and minimize the impact on customers' daily work. • Still, it was found that the customer involvement parts of the consolidation and ranking of standards (Steps 1 and 2) should only be combined if there are no likely changes to the standards discussed, otherwise the rankings may turn out to be obsolete and have to be repeated, resulting in extra efforts and, worse, customer annoyance.
Measurements (part of customer involvement in Steps 1, 2 and 3)	• Not many customers bothered to even open a questionnaire that was *attached* to an e-mail sent to them (as a proper attachment that can not be read without opening it), let alone fill it out and send it back.

Topics	Lessons learned
	• It proved to be far more promising to send the questionnaire as part of the e-mail, so that customers could directly see it without having to open an attachment (with the e-mail software used by the company it was possible to read the content of an e-mail – but not that of any attachments – without actually opening it).
	• Both English and the local language should be used to write the questionnaire. This fact should be clearly indicated using little flags that also serve as an eye-catcher.
	• The response rate to questionnaires in the case study was just under 20 per cent, which is quite high for questionnaires and seems to underline the effectiveness of talking to customers about the importance of their participation before sending out the questionnaires.
	• No responses were received anonymously via fax or letter, but all were sent back via e-mail. If there actually were customers wishing to give anonymous feedback they did not make the effort of printing the questionnaire, filling it out and sending it back by letter or fax.
	• A very small number of responses containing destructive criticism were received that could be addressed directly by contacting the respective customer. Reasons for destructive criticism were usually that some problem of a customer had not been properly addressed by the service provider, or was not even of the responsibility of the service provider. In any case, it proved to be very useful to directly contact such customers: Either the problem could then be properly addressed, or the customer could be told that the service provider could not do anything about it because the problem was outside the scope that the service provider could influence. In both cases, there are good chances that the angry customer could be turned around.
	• Feedback to questionnaires usually arrived within two weeks.
	• The response rate could be slightly improved by talking to customers about the questionnaire first, telling them that it would be important if they could participate.
	• Low response rates to questionnaires can be compensated by individual interviews and/or focus groups. In the case study, individual interviews were shown to be a good compensation for the lack of participation amongst customers.
	• Results from questionnaires, interviews and focus groups did not show significant differences between individual customer teams so that the average values were used to visualize both importance of and performance against service standards.
Internal resistance (by individual staff members of the service provider)	• In the given case study, there was no open internal resistance against the service quality cycle, or against concrete action items derived from using the method. What was found to start with was certain scepticism amongst the more experienced members

Topics	Lessons learned
	of the team, especially before the first results of measurements were available.
	• It proved to be very useful to allow open discussions within the team about the method and how it could be applied. Since all members of staff were very open-minded, although critical, all soon understood the advantages of the method and, as soon as the first measurement results were available, even initial doubters were convinced of both the effectiveness and the efficiency of the method.
External resistance (by individual customers)	• There was no open external resistance from customers against the implementation of the method, because the only thing they perceived were the specific customer involvement sub-steps used to define and rank the new service standards, as well as to measure performance against those standards. However, strong criticism was received from a few individual customers initially; for example, because they were expected to spend a couple of minutes to fill in a questionnaire. After explaining to such customers why it was important, they usually stopped complaining, but simply did not react to any questionnaires.
	• After all, there will always be some customers that wish to do everything the way it 'always' has been done. The way to deal with those customers, in the case study, was to try and fulfil their needs as much as they let the service provider, to keep offering to talk to them, to send them the reports and questionnaires (as to any other customer) but not to be surprised if they did not respond.
Statistical validity of measurements	• The number of participants in questionnaires was small, participants were not always the same, and a mix of means to gain customer feedback was used (questionnaires, interviews and focus groups) so that measurement results cannot be considered to be statistically valid. They offer, however, valuable indications and hints that are argued to be sufficiently accurate to effectively and efficiently support decision-making and improve and control customer-perceived service quality over time.
Effectiveness of applying the service quality cycle	• The application of the service quality cycle as proposed in Chapter 3 and described in the present case study has proven to be *highly effective*. Customer-perceived performance of the service provider against customer-driven service quality standards could be improved significantly, stakeholders in the service provider could be actually shown the levels of service quality achieved, and a sound basis for operational and strategic decision-making was generated that was not previously available. Since the application of the method will lead to improved levels of service quality and those improvements are presented in visible form, members of staff and management alike can and should be shown the results achieved, contributing

Topics	Lessons learned
	to higher levels of motivation amongst team members and recognition by management.
Efficiency of applying the service quality cycle	• The application of the service quality cycle has also proven to be *very efficient*. Only limited efforts were needed to implement the method (see Table 4.13). Once the cycle was closed for the first time, there was hardly any effort needed at all to measure performance and to re-present it visually it on a regular basis. Talks with customers had to be held anyway, only that when using the method those talks (interviews and/or focus groups) could be used in a more structured way to gain the information needed.

- *highly effective*; that is, customer-perceived performance of the service provider against customer-driven service quality standards could be improved significantly and a sound basis for operational and strategic decision-making was generated, which was not previously available
- *very efficient*; that is, only limited efforts were needed to implement the method and thereby achieve high effectiveness.

The advantages of using the proposed method may be compelling or seem logical but it is very difficult and possibly not quite honest to express them in a clear-cut amount of money saved. This is so for a number of reasons. For instance, all advantages may not be entirely caused by the use of the proposed method but partly by the behaviour or work of individual members of the service provider or customers. Also, the estimation of money saved per advantage is highly subjective and arbitrary. Nevertheless, this estimation of saving potentials is likely to be just what decision-makers want you to produce, so that they can base their judgement on those figures.

What is far easier to quantify, however, are the costs of implementing the proposed method. In the given case, costs have been calculated at a fixed rate of €70 per hour. Table 4.13 shows the non-reoccurring costs of implementing the service quality cycle in the case study, excluding the working time of customers (for instance when participating in interviews), but including working time of members of the service provider (such as participating in an internal workshop to derive action items).

When repeating any of the steps of the cycle at specific time intervals, only very little extra effort was spent in connection with the method; for example, to measure performance again, analyse the situation and derive action items.

Compared to the 2005 budget in the case study (€640,000), the implementation costs represented less than 0.7 per cent of the budget allocated to the service provider. To repeat specific steps at certain time intervals only took the corresponding time of those steps (as stated in Table 4.13).

Now that this chapter has presented an example implementation of the method proposed in Chapter 3 in the form of a recent case study from the European aerospace

Table 4.13   Costs of implementing the service quality cycle

Steps	Sub-activities	Hours	Costs (€)
1. Generation of service standards	Internal definition of service standards	8	560
	Interviews/focus groups	6	420
2 Ranking of service standards	Internal ranking of service standards	4	280
	Interviews/focus groups	6	420
3 Measuring performance	Preparation of questionnaire	4	280
	Interviews/focus groups	5	350
4 Analysing the current situation	Visualization of results	4	280
	Analysis	4	280
5 Deriving action items		18	1,260
6 Closing the cycle		4	280
Total non-reoccurring costs			**4,410**

industry, the following chapter will present general conclusions from the theoretical work and practical experience that were discussed in this book.

Chapter 5

Conclusion

After looking, in the previous chapter, at a concrete example implementation of the service quality cycle by an internal service provider within a leading aircraft manufacturer, the purpose of this chapter is to put the present book into the context in which it was written, and discuss some future steps in further refining the proposed method. The latter can be achieved by gaining new insights and experience from implementations in other aviation and comparable non-aviation domains that are conducted by both external and internal service providers.

There is a rich body of generally accepted knowledge and many theories available in the fields of 'services marketing', 'management of operations' and 'change management', some of which have been considered and reviewed from an aviation perspective in Chapter 2, to gain an overview and better understand the bases of the method suggested in Chapter 3. The proposed service quality cycle does not, however, incorporate entirely all of the theories discussed as such. Rather, it uses individual aspects and frameworks or tools to form a more practical and holistic approach, which is 'in reach' of service providers in the aviation context, even for smaller and/or internal service providers.

The works of Zeithaml and Bitner, as well as Slack, Chambers and Johnston that were drawn from and reviewed in Chapter 2 of this book are examples of very comprehensive, well researched theories and frameworks that are most valuable for any service provider and that are already (in parts) widely used in certain service-driven operations. Also, in some service-driven markets, there already is a lot of practical experience available and best practices in providing services in different domains are continuously established, challenged and improved. Some examples are leading airlines, many airport services providers, as well as customer support services delivered by aircraft manufacturers to customer airlines.

Still, there are many areas that have not originally or traditionally been considered to be service-driven such as, for instance, engineering and engineering support functions within aircraft manufacturing companies or suppliers and sub-contractors of such companies. In those organizations this knowledge and experience is not currently used to the extent it could be, leaving a lot of potential for improvements of service quality, with all the related benefits for all partners concerned.

The prevailing attitude is still very often that only specifications, documents or parts have to be delivered, not services. This lack of service culture awareness is (more often than not) dominant, even in leading aerospace manufacturing companies, not usually externally towards the end customers, but internally. Internal service culture is rarely heard of; even less often actually lived in such companies. Moreover, when people talk about service quality, they usually mean external service quality as perceived by the external customer. Internal service providers who apply proactive service quality management are, unfortunately, still the rare exception.

As a result, processes inside companies, lead times, internal costs, and the quality of such companies' end products (whatever those may be), not to mention motivation and mutual trust amongst employees, are all worse than they could be. From a financial perspective, this all comes down to those companies spending more money than needed to produce and market their products and services; that is, they are wasting resources. In other words, they could do better in their marketplace with their existing resources.

To summarize, it can be said that there remains a lot to do – especially for internal service providers – in order to materialize the potential for saving resources and/or improving the products and services they deliver.

This book is an attempt to build on some of the most recognized theories and frameworks to form one easy-to-use yet highly effective practical guide for both external and internal service providers in the aviation context, with the main focus on airline, airport and aircraft manufacturing operations. In particular, those parts of the industry are targeted that have not previously and/or traditionally been considered as service-driven. The suggested method is the service quality cycle, that offers a step-by-step approach to systematically control and improve service quality as perceived by customers. Although primarily intended for use in the aviation industries or related industries, it would equally be very beneficial in other domains not mentioned in the book. However, some adaptations may be necessary.

The case study presented in the previous chapter was about an example implementation of the proposed method by an internal service provider within an aircraft manufacturing company. In the near future, it would be interesting to follow up other implementations of this method, both with external service providers in the aviation industries, and with internal and external service providers in other industries or markets not directly related, in which the service aspect is (or perhaps should be) considered to be of high importance.

Having looked at the context in which this book was written, as well as some future steps that should be taken to enable further refinement of the proposed method over time, the following pages offer valuable sources for more detailed readings in the form of references mainly from the fields of services marketing, management of operations and change management that were used in the second chapter of this book.

The appendix offers a summary of all the step-by-step checklists from the method suggested in Chapter 3: these can be used as implementation checklists which can easily be adjusted to the individual needs and circumstances of the user.

Appendix

The Service Quality Cycle Checklists

Step 1

Table 3.1   Checklist - marketing research

- Actively find out what are reasons for customer complaints.
- Record those reasons.
- Actively find out why customers have ceased using the services.
- Record those reasons.
- Actively find out the best practices in delivering the services offered.
- Record those best practices.
- Actively find out present and future customer requirements.
- Record those customer requirements.

Table 3.2   Checklist – customer relationship management

- Monitor relationships with key customers.
- Keep talking with customers and own contact employees.
- Track and anticipate recovery opportunities.
- Take care of customer problems on the front line.
- Solve problems quickly.
- Empower front-line employees to solve problems.
- Record and learn from recovery experiences.
- Honestly appreciate and respect each individual customer and show it.

Table 3.8   Checklist – performance objectives (internal)

- Use the framework of the five performance objectives (quality, speed, dependability, flexibility and cost) to define a set of relevant and realistic performance objectives.
- Check that those performance objectives are in line with the overall strategy.
- Check that those performance objectives reflect what is perceived by front-line employees to be customer needs and requirements.
- Define partial measures for each of the objectives identified.
- Check that those partial measures are easy to measure, represent a mix of hard and soft measures, reflect assumed customer needs, and can actually be influenced by the service operation in question.
- Translate the above into a set of challenging yet realistic service standards that are easy to understand by customers and employees alike.

Table 3.9   Checklist – interviews and focus groups

- Select the means of customer involvement (interviews, focus groups).
- Prepare a structure for interviews and/or focus groups that is based on the service standards and the selected scale of importance.

Table 3.10   Checklist – customer involvement (Step 1)

- Identify suitable (existing and potential) key customers and possibly select a smaller list of candidates to be asked for their involvement.
- Invite the selected candidates, explaining the purpose of the meeting and underlining the importance of their help for your ability to improve the services delivered to customers.
- During the meetings use the set of service standards as a structure within which to retain and note all relevant inputs (comments, emotional reactions, etc.).

Table 3.11   Checklist – service standards (external)

- Collect all comments and other inputs made by the participants during the interviews and/or focus group meetings.
- Group those for each service standard that was discussed.
- Decide which of those inputs are justified and can or should be taken into account.
- If possible, use key employees to help you make that choice.
- Adjust the set of service standards accordingly.

Step 2

Table 3.12   Checklist – scale of importance

- Select a suitable scale of importance.
- Check the scale is sufficiently detailed to allow for clear ranking of service standards, yet immediately understandable for customers.

Table 3.13   Checklist – interviews, focus groups and questionnaires

- Select the means of customer involvement (interviews, questionnaires, focus groups).
- If interviews and/or focus groups are used, prepare a structure for those (based on the service standards and the selected scale of importance).
- If a questionnaire is used (based on the service standards and the selected scale of importance), formulate it carefully and as clearly as possible to limit the room for different interpretations.
- Where appropriate, prepare two versions of the questionnaire, one in the local language and one in English.
- Check the questionnaire is clear, short and simple.

Table 3.14   Checklist – customer involvement (Step 2)

- If interviews and/or focus groups are used, identify suitable (existing and potential) key customers and possibly select a smaller list of candidates to be asked for their involvement.
- Invite the selected candidates explaining the purpose of the meeting and underlining the importance of their help for your ability to improve services delivered to customers.
- During the meetings use the set of service standards as a structure within which to retain and note all relevant inputs (comments, emotional reactions, etc.).
- If questionnaires are used, send them out to all or specific customers.
- Tell customers you meet during normal service encounters that a short questionnaire is going to be sent out to them and that it would be very helpful if they could participate (this is likely to increase the reply rate).

Table 3.15   Checklist – table of ranking results

- Collect all replies and other inputs made by participating customers.
- Group them per service standard.
- Present the ranking choices in a spreadsheet (table).
- If there are big differences between specific customers' rankings, highlight those differences.
- If almost all customers show the same ranking behaviour, use the average ranking of the service standards.

Step 3

Table 3.16   Checklist – scale of performance

- Depending on the circumstances, decide whether the scale of performance should either reflect performance as compared to competitors (profit centre, own business, etc.) or very poor versus very good performance (cost centre, internal service operation, etc.).
- Select a suitable scale of performance.
- Check that the scale is sufficiently detailed to allow for meaningful performance measurements against the declared service standards.
- Check that the scale is immediately understandable for customers.

Table 3.13   Checklist – interviews, focus groups and questionnaires

- Select the means of customer involvement (interviews, questionnaires, focus groups).
- If interviews and/or focus groups are used, prepare a structure for those (based on the service standards and the selected scale of performance).
- If a questionnaire is used (based on the service standards and the selected scale of importance), formulate it carefully and as clearly as possible to limit the room for different interpretations.
- Where appropriate, prepare two versions of the questionnaire, one in the local language and one in English.
- Check the questionnaire is clear, short and simple.

Table 3.17   Checklist – customer involvement (Step 3)

- If interviews and/or focus groups are used, identify suitable (existing and potential) key customers and possibly select a smaller list of candidates to be asked for their involvement.
- Invite the selected candidates, explaining the purpose of the meeting and underlining the importance of their help for our ability to improve our services delivered to them.
- During the meetings use the set of service standards as a structure within which to retain and note all relevant inputs (comments, emotional reactions, etc.).
- Take note of both types of information, not only the measurement values per service standard, but also any other background information as to why specific customers came up with a given value.
- If questionnaires are used, send them out to all or specific customers.
- Tell customers you meet during normal service encounters that a short questionnaire is going to be sent out to them and that it would be very helpful if they could participate (this is likely to increase the reply rate).

Table 3.18   Checklist – table of performance results

- Collect all replies and other inputs made by participating customers.
- Group them per service standard.
- Present the performance measurements in a spreadsheet (table).
- If there are big differences between specific customers' perceptions, highlight those differences.

Step 4

Table 3.19   Checklist – importance-performance matrix

- Draw a generic importance-performance matrix.
- Use the scale of importance that was selected during Step 2 on the x axis.
- Use the scale of performance that was selected during Step 3 on the y axis.
- Select a recognizable symbol for each standard in the matrix (such as a small circle with the number of the service standard from the whole set of selected standards).
- Look at each service standard individually and insert the corresponding symbol into the matrix, using the ranking value on the x axis and the rating value on the y axis.

Table 3.20   Checklist – situational context

- Consider the overall situational context of the service operation and its customers that has not directly been influenced by either the service operation or its customers.
- Consider internal factors that have been influenced by the service operation; that is, its employees (such as recent changes in the scope of services or how they are delivered to customers, concrete actions to improve customer-perceived performance, etc.).
- Consider external factors that have been influenced by customers (such as any recent changes in using the services offered).

Table 3.21   Checklist – analysis

- Briefly describe the overall situational context that cannot be directly influenced by the service operation, nor by its customers.
- Interpret each service standard individually.
- Where does it stand in the matrix and what does this mean?
- What internal reasons might there be for the current position?
- What external reasons might there be for the current position?
- Is that position acceptable under the circumstances (given context)?
- If not, where would be an acceptable position in the matrix?
- What are the alternative options to shift its position in the matrix?
- What are the advantages and drawbacks of each option?

Step 5

Table 3.22   Checklist – strategy

• What are the overall short-, medium- and long-term strategic goals of the customer organizations?
• What are the overall short-, medium- and long-term strategic goals of the service operation?
• What is the expanded marketing mix of the service operation (product/service, place, promotion, price, people, physical evidence, process)?
• Going through all service standards one by one, which of the options presented in the analysis are in line with the customers' and the service operation's strategic goals and fit the current expanded marketing mix of the service operation or could be adjusted to fit?

Table 3.23   Checklist – contractual obligations

• What are the current contractual obligations the service operation has to fulfil?
• Going through all service standards one by one, which of the options presented in the analysis are ruled out by those contractual obligations?

Table 3.24   Checklist – infrastructure

• What is the current infrastructure able to deliver the services to customers and what are the infrastructural limitations?
• Could any of these infrastructural limitations be overcome and how?
• Going through all service standards one by one, which of the options presented in the analysis are ruled out by current infrastructural limitations that are not likely to be overcome in the future?

Table 3.25   Checklist – financial resources

• What currently are the financial resources available to the service operation and what are those predicted to be in the future?
• Going through the service standards one by one, which of the options presented in the analysis are ruled out by lack of current and/or predicted financial resources?

Table 3.26   Checklist – human resources

- What are the current human resources of the service operation?
- What are the qualifications, potential, experience and motivation of these people?
- Have you got the resources and/or authority to recruit additional team members?
- Going through the service standards one by one, which of the options presented in the analysis are ruled out by lack of suitable personnel?
- Could this lack be compensated for by hiring new people and/or training the existing team members?

Table 3.27   Checklist – action items

- Going through the service standards one by one, which of the options presented in the analysis have not been ruled out by one of the previous considerations?
- Which of those or combination of those seem to be the best choice under the circumstances?
- For service standards where all options have been ruled out, is there a compromise solution that limits the damage to the customer and the service operation?
- Generate a list of best-choice concrete action items based on the previous considerations.
- State the desired effects and risks of each of those action items.

Step 6

Table 3.28   Checklist – dealing with resistance

- Consider what internal resistance could be expected.
- Keep your team sufficiently involved and informed along the entire service quality cycle.
- Take time to convince them, openly listen to their ideas, worries, doubts or fears, and seriously take them on board.
- Consider what external resistance could be expected.
- Identify which customers are influencers and/or opinion leaders and/or could be used as change agents.
- In the case of major changes, choose a suitable implementation strategy.
- Make sure your team members are sufficiently empowered.

Table 3.29   Checklist – priority

- Decide on the priority of individual action items.
- If there are not sufficient resources to implement all action items at once, decide which should be addressed first, based on their priority.

Table 3.30   Checklist – timing

- Consider when the resources that are needed to implement each action item will be available.
- Consider the customers' schedules, especially if their direct involvement is needed to implement a specific action item.
- Consider when there are relevant customer deadlines and/or milestones.
- Based on the above considerations and the priority allocated to each action item, schedule when each one of them will be implemented.

Table 3.31   Checklist – communications to customers

- Proactively influence customer *expectations* by making realistic service promises (both explicit and implicit), educating customers about their role, identifying and targeting influencers and opinion leaders, as well as telling customers when service provision is higher than what could normally be expected.
- Proactively influence customer *perceptions* by aiming for customer satisfaction in every service encounter, actively managing service evidence, and using customer experiences to reinforce images.
- Exceed customer expectations by under-promising and over-delivering, as well as positioning unusual service as unique, not the standard.
- All communications must be in line with the service standards selected.

Table 3.32   Checklist – closing the cycle

- Based on the above considerations, allocate each action item to one or several persons who are responsible for carrying it out.
- Clearly state any steps needed to pre-empt internal or external resistance.
- Going through the list of action items decided, clearly describe the timeframe for each action, its priority, and how it will be communicated internally and externally.
- For each action item, decide how it will be followed up over time to ensure that the expected improvements really materialize.
- Decide in what time intervals the service quality cycle should be repeated and whether the customer involvement parts of Steps 1, 2 and 3 could be combined in some form, in order to minimize disturbance for customers, once the cycle has been established.

Bibliography

Brancheau, J.C., and Wetherbe, J.C. (1994), 'Understanding Innovation Diffusion Helps Boost Acceptance Rates of New Technology', in Gray, P., King, W.R., McLean, E.R., and Watson, H.J. (1994), *Management of Information Systems* (Orlando FL: The Dryden Press).

Daniels, C. (1998), *Information Technology – The Management Challenge* (Harlow: Addison-Wesley).

De Vicuña Ancín, J.M.S. (1999), *El Plan de Marketing en la Practica* (Madrid: Editorial ESIC).

Dibb, S., Simkin, L., Pride, W.M., and Ferrell, O.C. (1997), *Marketing Concepts and Strategies* (Boston: Houghton Mifflin).

Engelhardt, W. (1996), 'Das Service-optimum im Spannungsfeld von Kundenzufriedenheit und Effizienz', in Kleinaltenkamp, M., Fließ, S., and Jakob, F. (1996), *Customer Integration – Von der Kundenorientierung zu Kundenintegration* (Wiesbaden: Gabler).

Goleman, D. (1999), *Working with Emotional Intelligence* (London: Bloomsbury Publishing).

Gray, P., King, W.R., McLean, E.R., and Watson, H.J. (1994), *Management of Information Systems* (Orlando FL: The Dryden Press).

Günter, B., and Huber, O. (1996), 'Beschwerdemanagement', in Kleinaltenkamp, M., Fließ, S., and Jakob, F. (1996), *Customer Integration – Von der Kundenorientierung zu Kundenintegration* (Wiesbaden: Gabler).

Hammer, M., and Stanton, S. (2000), 'Prozessunternehmen – wie sie wirklich funktionieren', *Harvard Businessmanager* 3/2000, 68.

Hope, C., and Mühlemann, A. (1997), *Service Operations Management – Strategy, Design and Delivery* (Hemel Hempstead: Prentice Hall Europe).

Kleinaltenkamp, M., Fließ, S., and Jakob, F. (1996), *Customer Integration – Von der Kundenorientierung zu Kundenintegration* (Wiesbaden: Gabler).

Kotler, P., Armstrong, G., Saunders, J., and Wong, V. (1996), *Principles of Marketing – The European Edition* (Hemel Hempstead: Prentice Hall Europe).

Kotter, J. (1996), *Leading Change* (Boston: Harvard Business School Press).

Kotter, J., Schlesinger, L.A., and Sathe, V. (1986), *Organization: On the Management of Organizational Design and Change* (Homewood IL: Irwin).

Palmer, A. (1994), *Principles of Services Marketing* (Maidenhead: McGraw-Hill).

Santesmases, M. (1994), *El Marketing Financiero como Factor de Competitividad* (Madrid: Papeles de Economía Española No. 58).

Slack, N., Chambers, A., and Johnston, R. (2004), *Operations Management* (London: Pearson Education).

Storey, J. (1995), *Human Resources Management* (London: Routledge).

Zeithaml, V.A., and Bitner, M.J. (2002), *Services Marketing* (Singapore: McGraw-Hill).
Zeithaml, V.A., Berry, L.L., and Parasuraman, A. (1993), 'The Nature and Determinants of Customer Expectations of Service', *Journal of the Academy of Marketing Science* 21, 1.

Websites:

www.airbus.com
www.airlinequality.com
www.baldrige.com
www.btre.gov.au
www.bts.gov
www.faa.gov
www.fraport.de
www.iata.org
www.iso.org
www.munich-airport.de
www.quality.nist.gov
www.quality.org
www.singaporeair.com
www.skytraxsurveys.com
www.whitehouse.gov

Index

action items, service quality cycle 111-17
'adequate service' 23
Airbus *26-7*, 57
Aircraft Component Management Team
    (ACMT) 132
aircraft manufacturing 7-8, 28-9
    Airbus *26-7*, 57
    service quality cycle *91, 97, 103, 110,*
        117, 125
    see also Support Services Group
        (aircraft manufacturing case study)
airlines 8-9
    dba *41*
    easyJet *37*
    Niki 15-16
    PrivatAir *19*
    rankings (SKYTRAX) 69-70, *71, 72*
    Ryanair *56-7*
    service quality cycle *91, 97, 103, 110,*
        117
    Singapore *73*
airports 9-11
    Fraport AG *32-3*
    Luebeck *46-7*
    Munich *66-7*
    rankings (SKYTRAX) 70
    service quality cycle *91, 97, 103, 110,*
        117, 125
analysis, service quality cycle 102-11
Available Seat Kilometres (ASK) 6, *7*
awards and prizes 65

Baldridge National Quality Award 65
Brancheau, J.C. and Wetherbe, J.C. 59

cabin crew 4
capability
    aircraft manufacturing case study 129,
        130, 133, *134*
    definition xiii

capacity
    and demand factors 39-40
    growth, passenger traffic and 6, *7*
'cargo' carriers 9
change
    management issues 55-62
    resistance to 58-9, 118-21, *125*
cleaning services 4-5
client *see entries beginning* customer
closing the cycle, service quality cycle
    117-26
coaching, definition xiii
communication(s)
    matching performance to promises 42,
        43-5
    to customers 121-3
    IT service providers 43-5, *82, 83, 84-5*
complaint management 25-8, 29
Component Design Build Team (CDBT)
    132
Component Management Integration Team
    (CMIT) 132
contractual obligations, service quality cycle
    113
cost
    as performance objective 49-50, *86*
    see also financial issues; low-cost
        airlines
customer expectations 21-3
    provider knowledge of 28-31
    see also importance-performance matrix
customer gaps 21-8
customer integration 35-6
customer involvement 88-9, 94-5, 100-101
    see also service quality cycle (steps 1-3)
customer perceptions 23-5, 123
customer relations building 29, 30-31
customer relationship marketing 30, 79-81

customer(s)
  aircraft manufacturing case study 131-2,
    133, *134*
  communications to 121-3
  definition xiii-xiv
  factors in non-delivery of service
    standards 39
  retention strategies 30
  word-of-mouth reports 79-80

Daniels, C. 55-7
dba *41*
De Vicuna Acín, J.M.S. 14
demand and capacity factors 39-40
dependability, as performance objective
    48-9, *83*, *86*
Design Build Team (DBT) xiv, 132
designs, provider gap 31-6
'desired service' 22-3
Dibb, S. et al. xiv, 29

easyJet *37*
employees 38-9, 89
  cabin crew 4
  flight crew/attendants 4, 107-10
  front-line 30, 31, 80
  see also human resources
Engelhardt, W. 14
evidence of service 23
expanded marketing mix 16-18, *117*
external customer, definition xiv

financial issues
  aircraft manufacturing case study 131
  cost as performance objective 49-50, *86*
  incentive for service improvement 2-3
  price/pricing 24, 42
  service quality cycle 114
five performance objectives *see* performance
    objectives
flexibility, as performance objective 49,
    84-5, *86*
flight crew/attendants 4, 107-10
focus groups 86-7, 93-4, 100
  definition xiv
France 131-2, 133
Fraport AG *32-3*
front-line employees 30, 31, 80

gaps model *see* customer gaps; provider
    gaps
Germany 9-11, 130-31, 133
  airports *32-3*, *46-7*, *66-7*
Goleman, D. 60
Gray, P. et al. 59
Günter, B. and Huber, O. 25-8

Hammer, M. and Stanton, S. 58
help desk support, definition xiv
Hope, C. and Mühlemann, A. 35
human resources
  aircraft manufacturing case study 131
  service quality cycle 115, *116*
  see also employees

image of service 24
importance of providing service quality 5-6
importance-performance matrix 51-5, 105-6,
    107
information technology *see* IT service
    providers
infrastructure, service quality cycle 114
intermediaries, in non-delivery of service
    standards 39
internal customer, definition xiii
internal service providers 3-6
International Air Transport Association
    (IATA) 6
international marketing 40
interviews 86-7, 93-4, 100
  definition xiv
ISO 9000 Series 64-5
IT service providers 43-5, *82*, *83*, *84-5*

Kleinaltenkamp, M., Fleiß, S. and Jakob,
    F. 35-6
Kotler, P. et al. 29
Kotter, J.P. 60-62
Schlesinger, L.A. and Sathe, V. 58-9

Lauda, Niki (Niki airline) 15-16
leadership 36, 60-62
low-cost airlines 8-9
  definition xiv
  vs. traditional airline 53-5
Luebeck Airport *46-7*

management
    aircraft manufacturing case study 129, 130
    and leadership distinction 62
market for service providers 6-11
marketing
    customer relationship 30, 79-81
    definition xiv
    expanded mix 16-18, *117*
    international 40
    research 29, 78-9
Munich Airport *66-7*

new technology *see* IT service providers
Niki (airline) 15-16

operations, definitions xv, 14-15

Palmer, A. 14
passenger traffic and capacity growth 6, *7*
perceived service 23-5, 123
performance objectives 45-51
    importance-performance matrix 51-5
    internal 81-6
performance results table 101-2, 105
physical evidence, matching performance to promises 42
positioning, definitions xv, 15-16
price/pricing 24, 42
priority, service quality cycle 121, *185*
PrivatAir *19*
prizes and awards 65
provider gaps 20, 21
    1 - not knowing what customers expect 28-31
    2 - not selecting right service designs and standards 31-6
    3 - not delivering to service standards 38-40
    4 - not matching performance to promises 40-45

qualitative and quantitative standards 34
quality
    in aviation 62-74
    as performance objective 48, *82, 86*
    see also service quality cycle

quality management systems
    benefits of 63
    ISO 9000 Series 64-5
questionnaires 93-4, 100
    definition xv

rankings 68-9
    airlines 69-70, *71, 72*
    airports 70
    customer involvement results 95, *96*, 105
    definition xv
    of service standards 90-102
recovery plans 30-31, 80, *91*
relationship marketing *see* customer relationship marketing
reports
    Air Travel Consumer Report 74
    word-of-mouth 79-80
resistance to change 58-9, 118-21, *125*
retention strategies 30
Revenue Passenger Kilometres (RPK) 6, *7*
Ryanair *56-7*

safety, definition xv
Santesmases, M. 14
scale of importance, service standards 93
scale of performance, service standards 99-100
security, definition xv
security services 5, 51, *82, 83, 84*
service, definitions and properties xv, 14-15
service encounters 23
service leadership *see* leadership
service operations, definitions xv, 15
service positioning *see* positioning
service providers
    definition xv
    market for 6-11
    provision by 3-6
    see also provider gaps
service quality cycle 75-6
    checklists 177-86
    step 1 - generation of service standards 76-90
    step 2 - ranking of service standards 90-102

step 3 - measuring performance against
    service standards 96-102
step 4 - analysing current situation
    102-11
step 5 - deriving action items 111-17
step 6 - closing the cycle 117-26
summary of benefits 126-7
see also Support Services Group
    (aircraft manufacturing case study)
service standards
    definition xv-xvi
    external 89-90, 92-3, 98
    provider gap 31-40
    see also service quality cycle (steps 1-3)
Singapore Airlines *73*
situational context 106-7
SKYTRAX rankings 68-70
Slack, N., Chambers, A. and Johnston, R.
    14-15, 45, 50, 51, 175
spare-parts dispatch unit 50, *82, 83, 85*
speed, as performance objective 48, *83, 86*
standards
    ISO 9000 Series 64-5
    see also service standards
Storey, J. 60
strategies, service quality cycle 113
sub-contractors
    of aircraft manufacturers 8
    definition xvi
suppliers
    of aircraft manufacturers 8
    of airline operations 9
    of airport operations 11
    definition xvi

support services, definition xvi
Support Services Group (aircraft
    manufacturing case study)
    capability (teams) 129, 130, 133, *134*
    context 129-32
    customers 131-2, 133, *134*
    financial resources 131
    human resources 131
    management 129, 130
    program support teams 129, 130
    service provider roles 130-31
    service quality cycle 133-66
        lessons learnt 169-74
        post-implementation status 166-9
        pre-implementation status 133
        step 1 134, 135-47
        step 2 134, 147-51
        step 3 134, 151-6
        step 4 135, 156-66
systems thinking 55-7

technical support and maintenance 4
theoretical considerations 13-18
    gaps model *see* customer gaps; provider
    gaps
timing, service quality cycle 121, *185*
training, definition xvi

United States (US) 9, 65, 74

word-of-mouth reports 79-80

Zeithaml, V.A. and Bitner, M.J. 2-3, 15, 16-
    17, 20, 22, 29, 30, 115, 175